Ríos de Anda / Sotta

Sorption and Diffusion in Polymers

Agustín Ríos de Anda
Paul Sotta

Sorption and Diffusion in Polymers

Functional Materials and Applications

HANSER

Print-ISBN: 978-1-56990-793-1
E-Book-ISBN: 978-1-56990-794-8

Bibliographic information of the German National Library:
The German National Library lists this publication in the German National Bibliography; detailed bibliographic data are available on the Internet at http://dnb.d-nb.de.

Vilshofener Straße 10 | 81679 Munich | info@hanser.de
www.hanserpublications.com
www.hanser-fachbuch.de
Editor: Dr. Mark Smith
Production Management: Cornelia Speckmaier
Cover concept: Marc Müller-Bremer, www.rebranding.de, Munich
Cover design: Max Kostopoulos
Cover picture: © Max Kostopoulos, unter Verwendung von chatgpt.com
Typesetting: Eberl & Koesel Studio, Kempten

The Authors

The Editors

Dr. Agustín Ríos de Anda

Dr. Agustín Ríos de Anda is a Professor at Ingénierie des Matériaux Polymères (Polymer Materials Engineering), a joint CNRS/Université Lyon/INSA Lyon lab in Lyon, France since September 2025. After obtaining his engineering degree in polymer chemistry in 2009 from the Institut National des Sciences Appliquées de Rouen, France, he undertook his PhD under Dr. Paul Sotta and Dr. Louise-Anne Fillot at the Advanced Polymers and Materials Laboratory in Lyon, France. His dissertation, presented in 2012, dealt with the influence of sorption of polar and non-polar molecules in nylon 6,6 and the consequences of such sorption and diffusion on the polymer's mechanical properties. He then conducted two postdoctoral periods of research at the University of Bayreuth, Germany, under Prof. Volker Altstädt, and at the Centre de Recherche sur les Macromolécules Végétales, Grenoble, France, under Dr. Laurent Heux. From September 2016 to August 2025 he held a tenure as Associate Professor at Université Paris Est Créteil. He has specialized in investigating the relationship between polymers' chemical and morphological structure with their macroscopic mechanical and transport properties through a multiscale experimental approach. For the past several years he has focused his research and this approach on bio-based functional polymers.

Dr. Paul Sotta

Dr. Paul Sotta is Senior CNRS Researcher at Ingénierie des Matériaux Polymères (Polymer Materials Engineering), a joint CNRS/Université Lyon/INSA Lyon lab in Lyon. He obtained his PhD in 1987 at the Université Paris-Sud (now Université Paris-Saclay), and then was a CNRS researcher at Laboratoire de Physique des Solides, a joint CNRS/Université Paris-Sud lab (1988–2006), invited researcher in the group of H. W. Spiess at the Max-Planck-Institut für Polymerforschung, Mainz, Germany (1995–1996), and CNRS senior researcher at the Laboratoire Polymères et Matériaux Avancés, a joint CNRS/Solvay lab in Lyon (2006–2020). He has made major scientific contributions in polymer physics, namely in local structure and dynamics in elastomers and reinforced elastomers, glass transition, reinforcement mechanisms, tear and wear resistance, polymer blends, deformation mechanisms in semicrystalline polymers, and solvent sorption in polymers. His fields of expertise cover both modeling and experimental techniques such as broadband dielectric spectroscopy, pseudo-solid NMR, X-ray and neutron scattering, tribology and contact mechanics, linear and non-linear rheology and mechanics.

The Contributors

Chapter No.	Title	Contributors
1	Modelling Solvent Sorption in Polar Polymers using a Physical, Isothermal, Non-Empirical Model	Agustín Rios de Anda, Paul Sotta
2	Influence of Organic Solvent Sorption upon Water Infiltration and Molecular Mobility in Polyamides	Agustín Ríos de Anda, Florentina-Maria Preda, Louise-Anne Fillot, Paul Sotta
3	Diffusion of Water and Oils in Thermosets and Thermoset Composites	Simon Bard, Martin Demleitner, Fabian Hübner, Volker Altstädt, Holger Ruckdäschel
4	Permeability of Oxygen through Polymers	Emmanuel Richaud, Alain Guinault
5	Water Sorption Behavior in Polymers: Kinetic and Thermodynamic Approaches	Stéphane Marais, Kateryna Fatyeyeva

Preface

One of the largest and most important applications of polymers has been, for several decades, their use as barrier materials. These applications range from the cellophane or polystyrene films used in food packaging to the use of technical polymers in oil and gas exploration and transportation. Among these numerous applications, the most common are the use of PET for liquid bottles and PE for milk bottles, PVC hoses for gardening and water circuits in buildings, polyamides in hoses at gas stations, polyfluorovinyls for raincoats and umbrellas, and polyvinyl alcohol for contact lenses, to mention a few. Novel barrier applications of polymers are found in the automotive industry where polyamides have been considered to be used in the fuel circuit (i. e. as gas tanks and also as the hoses connecting the tank with the engine). Other important current barrier/transport applications of polymers are as smart membranes for the controlled release of pharmaceuticals and as chemical sensors in medical applications. Further applications are found in the vast field of bladders of all kinds like in the tire industry, and sports applications (balls, equipment, etc.). Another key application is the large field of O-rings and sealings of all kinds, which are ubiquitous in vehicles as well as in gas, oil, and liquid transportation.

However, these barrier/transport properties are often neglected or relegated to the background behind their thermal, mechanical, and optical behavior. They are very often considered only from an engineering point of view. This means that tests performed to characterize the materials are engineering, normalized tests which deliver performance indices rather than well-defined physical quantities. This also means that most often empirical models are used to tentatively predict the performances of new materials by inter- or extrapolating the performance of known materials.

This book covers a theoretical part of transport phenomena, but focuses mostly on areas important to both academia and industry.

Several authors have studied these properties since the boom of such materials in the 1940s, the most well-known being Langmuir, Flory & Huggins, and Crank. Since then, several models of sorption of gas and/or liquid molecules in polymers have been proposed over the years to better understand and describe the various sorption mechanisms observed in polymers, especially in the case where molecules have strong affinities with the polymers and/or form so-called clusters within the materials. These models, which are presented in this book, describe well the sorption of solvents through a combination of a physical approach with empirical observations. These approaches yield qualitative parameters that are not able to completely describe the observed sorption phenomena physically. Thus, as an introductory part of this book a purely physical new model is proposed to describe the sorption of molecules having strong affinities with polymers possessing preferential sorption/interaction sites. This model yields physically meaningful parameters that describe the sorption of a given polymer/solvent system.

Subsequently in the book, transport phenomena in various applicative and functional polymers are discussed, with a focus on how to enhance such properties. Several areas relevant to the research and engineering fields on barrier/transport properties are thus considered, such as areas in which barrier properties are of direct, primary relevance in the application: fuel tanks and hoses, water and/or fluid circuitry, bladders, food packaging, and O-rings and sealings. While the barrier property is the primary concern, the final application often demands a compromise between barrier properties and other properties such as mechanical toughness (e. g. for fuel tanks), electrical insulation, etc., be achieved. Sometimes, optimizing both barrier properties and, for example, mechanical properties demands adjustment of some material feature in an antagonistic way. One example is the role of the glass transition temperature of the polymer. Decreasing solvent diffusion can be achieved by increasing the glass transition of the polymer, which means decreasing the molecular mobility at usage temperature, but this may make the polymer fragile, which is critical as regards its toughness.

This book aims thus to summarize and condense the latest research in the field of sorption, diffusion, and permeation in functional polymers.

A. Ríos de Anda, P. Sotta

July 2025

Contents

1 Modelling Solvent Sorption in Polar Polymers using a Physical, Isothermal, Non-Empirical Model

Agustín Ríos de Anda and Paul Sotta

1.1 Introduction

One of the largest and most important applications of polymers has been, for several decades, their use as barrier materials. These applications range from the cellophane or polystyrene films covering our food in supermarkets to the use of technical polymers in oil and gas exploration and transportation. Amongst these applications, the most common ones are the use of PET for liquid bottles and PE for milk bottles, PVC hoses for gardening and water circuits in buildings, polyamides in hoses at gas stations, polyfluorovinyls for raincoats and umbrellas, and polyvinyl alcohol for contact lenses, to mention a few. Novel barrier applications of polymers are found in the automotive industry, where polyamides have been considered for use in the fuel circuit (i. e. as gas tanks and also as the hoses connecting the tank with the engine) and technical metal-plated carbon-fiber-reinforced epoxides for hydrogen storage. Other important current barrier/transport applications of polymers are as smart membranes for the controlled release of pharmaceuticals, and as chemical sensors in medical applications.

However, these barrier/transport properties are often neglected or relegated to second place in front of their thermal, mechanical, and optical behavior. Nevertheless, several authors have studied these properties since the boom of these materials in the 1940s, the most well-known being Langmuir [1], Flory & Huggins [2–9], and Crank [10].

Since then, several models of sorption of gas and/or liquid molecules in polyamides have been proposed to better understand and describe the various sorption mechanisms observed in polymers, especially in the case where molecules have strong affinities with the polymers and/or form so-called clusters within the materials. These models, which will be presented and detailed in Section 1.2, describe well the sorption of solvents through a combination of a physical approach with empirical observations. However, these approaches yield qualitative parameters that are not able to completely physically describe the observed sorption phenomena.

To resolve this deficit, in Section 1.3 we describe a new, purely physical model describing the sorption of molecules with strong affinities to polymers that possess preferential sorption/interaction sites. We show that this model yields physically meaningful parameters that describe the sorption of a given polymer–solvent system. A comparison of the predictions of this model with experimental data is then provided in Section 1.4.

1.2 Sorption Mechanisms in Polymers

1.2.1 Sorption Thermodynamics

When a polymer is put into contact with a solvent under partial vapor pressure P, the solvent molecules penetrate into the polymer matrix up to a certain equilibrium concentration given by:

$$\mu_s(c, T) = \mu_{gas}(P, T) \tag{1.1}$$

Where:

$\mu_s(c, T)$ = chemical potential of sorbed molecules, which depends on the concentration c in the material and on T

$\mu_{gas}(P, T)$ = chemical potential of the gas, given by:

$$\mu_g(P, T) = RT \ln\left(\frac{P}{P_{sat}}\right) = RT \ln a \tag{1.2}$$

Where:

a = chemical activity

The chemical potential of the saturating vapor (gas at $P = P_{sat}$) is $\mu_g^{(0)}(T)$, and this vapor is in equilibrium with the liquid (thus $\mu_g^{(0)}(T) = \mu_{liq}(T)$), and has been chosen as the zero level of chemical potential. Then $\mu_{gas}(P, T)$ is negative if $P < P_{sat}$. For the gas phase at fixed pressure P, P/P_{sat} decreases as T increases. Thus, the chemical potential μ_{gas} decreases as T increases, and sorption decreases as T increases. The chemical potential of the liquid (which is nearly independent of P), which corresponds to a fixed activity $a = 1$, increases as T increases. Thus, in this case, the sorption increases as T increases. More generally, at a given, fixed activity $a = P/P_{sat}$ (which corresponds to a partial pressure of the gas P increasing as T increases), the chemical potential increases as T increases. Thus, the sorption increases as T increases.

The sorption thermodynamics is illustrated in Figure 1.1. Sorption isotherms, which relate the concentration *c* in the material to the pressure *P* or activity *a* of the vapor at a temperature *T*, are described by Equation 1.1. The chemical potential may be written $\mu_{ads}(c, T) = \Delta H - T\Delta S$, with an enthalpic contribution (enthalpy of sorption ΔH) and an entropic term ΔS, which depends on concentration *c*. Note that ΔH may possibly depend on temperature.

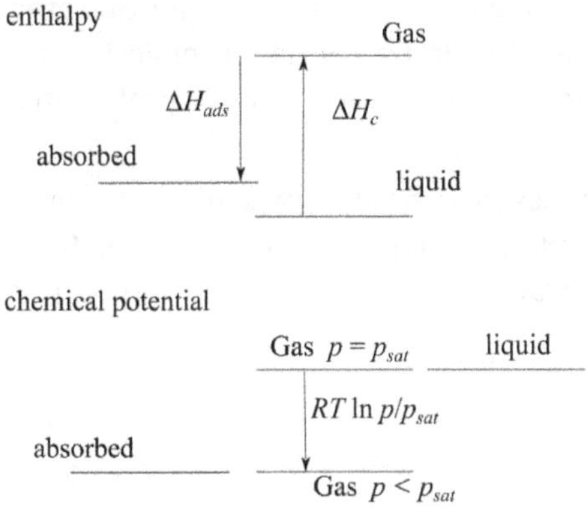

Figure 1.1
Thermodynamic relations related to sorption

By measuring sorption isotherms at various temperatures *T* and taking points at constant concentration *c*, the following curve is obtained [11]:

$$\ln a_c(T) = \frac{\Delta H_{st}}{RT} + \text{constant} \tag{1.3}$$

Then ΔH_{st} (at a given concentration *c*) can be obtained from the slope of the curve ln *ac* vs 1/*T*. This quantity is called the net isosteric heat of sorption, and is equal to $\Delta H_{st} = \Delta H_{abs} - \Delta H_{cond}$.

The sorption enthalpy $\Delta H_{abs} = H_{gas} - H_{abs}$ contains the contribution of molecular interactions of water molecules in an adsorbed state compared to water molecules in a gaseous state. In order to compare the interactions of water molecules sorbed in the polymer to those in liquid water, the heat of liquefaction ΔH_{cond} must be subtracted [11] (Figure 1.1). Note that Equation 1.3 is analogous to the Clapeyron equation for liquefaction [12] $\frac{\partial \ln P_{sat}}{\partial T} = \frac{\Delta H_{vap}}{T^2}$, where ΔH_{vap} is the enthalpy of vaporization.

1.2.2 Sorption Models

Several models have been developed in order to relate the equilibrium penetrant concentration to the penetrant activity, i. e., to describe the shape of sorption isotherms. The simplest one is Henry's law [13–15]. This model applies to the sorption of ideal gases at low activities, and states that at a constant temperature, the amount of gas that dissolves in a polymer (C_D) is directly proportional to the activity of that gas in equilibrium with the polymer:

$$C_D = K_D a \tag{1.4}$$

K_D is called Henry's constant and depends on the solute, the polymer and the temperature. This model applies to solutes that are sorbed at very low activity or that have a low solubility in the polymer matrix. In this case, K_D is equivalent to the solubility coefficient S. Henry's law is valid for any sorption mode at low solute activity. However, this model is not able to describe the sorption of molecules at high chemical activities and/or possible molecule–polymer interactions, or the occupation of preferential sorption sites by molecules.

The Langmuir sorption model [1, 16, 17] assumes there exists a fixed density of preferential sorption sites. It applies for instance to polymers containing specific functional groups, micro-voids, or high-area inorganic fillers. The sorption isotherm is described by:

$$\phi = \frac{a \exp(H/RT)}{1 + a \exp(H/RT)} \tag{1.5}$$

Where:

a = solvent activity

H = sorption enthalpy

ϕ = fraction of occupied sites in the material (coverage)

Corresponding sorption isotherms are concave with a rapid increase at low activities and a limiting coverage value (see Figure 1.2a). In the limit of low activity ($a \lll 1$), Equation 1.5 reduces to Henry's law with $\phi = a \exp(H/RT)$. Another way to consider Langmuir sorption is to equilibrate the rates of sorption/evaporation at equilibrium: writing rate of sorption $a(1 - \phi)$ equals the rate of evaporation $\phi \exp(-E/k_B T)$ gives Equation 1.5 as well.

The enthalpy H may be positive or negative. $H < 0$ corresponds to very small sorption. Sorption increases as H increases. For $H > 0$ (exothermic sorption), sorption of a molecule results in a gain in energy (because of attractive interactions (either non-polar such as van der Waals, or polar) developed with the surface, which are stronger than interactions within the liquid). The driving force for adsorption is enthalpic, with entropy acting against adsorption. It is thus expected that sorption will decrease as temperature increases (see Figure 1.2b). For $H < 0$ (endothermic sorption), molecules have to overcome an energy barrier in order to adsorb, and here the driving force for adsorption is entropy. Sorption will thus increase as temperature increases.

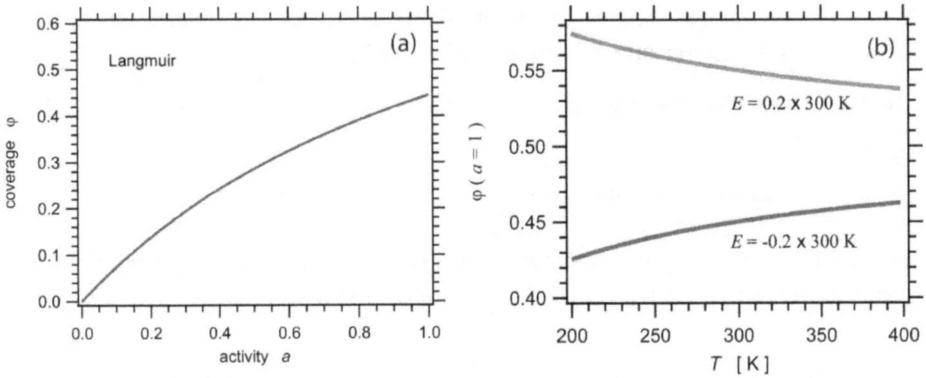

Figure 1.2 (a) Sorption isotherms in Langmuir sorption. The value $E/k_BT = -0.2$ was chosen. (b) Temperature variation of the sorption at equilibrium (activity $a = 1$) in Langmuir sorption. When E is negative, there is an energy barrier for sorption (endothermic sorption) and the driving force for sorption is entropy

Another model is the dual mode model [15, 18, 19, 20], which considers that for low molecule concentrations in the polymer, the sorption isotherm will be that of Henry's law. When the concentration starts to increase, the following absorbed molecules will occupy the free volume or preferential sorption sites, as described by Langmuir's sorption isotherm. The relationship describing the dual mode model is shown in Equation 1.6:

$$C = C_{\mathrm{D}} + C_{\mathrm{H}} = K_{\mathrm{D}}a + C'_{\mathrm{H}}\frac{ba}{1+ba} \tag{1.6}$$

Where:

C_{D} and C_{H} = Henry and Langmuir molecule concentration contributions

C'_{H} and b = Langmuir's capacity and affinity constant respectively

The Flory–Huggins sorption mode [2–9] is used to describe non-polar polymer–solvent mixtures. The affinity between the penetrant and the polymer matrix is described by an interaction parameter χ, a dimensionless measurement of the differences in interaction energies between species i and j in a mixture, which can be estimated with:

$$\chi = \frac{(\delta_i - \delta_j)^2}{RT}\sqrt{v_i v_j} \tag{1.7}$$

Where:

δ_i = Hildebrand solubility parameter

v_i = molar volume of species i (solvent or monomer)

Using the expression of the free energy of mixing between a very long polymer chain and a solvent, sorption isotherms are described by [2–9]:

$$\ln a = \ln \phi + (1 - \phi) + \chi(1 - \phi)^2 \tag{1.8}$$

Where:

ϕ = solvent volume fraction in the polymer

In the limit of small sorption, Equation 1.8 reduces to $\phi = a \exp(-(1 + \chi))$, which corresponds to Henry's law. A typical isotherm is shown in Figure 1.3a.

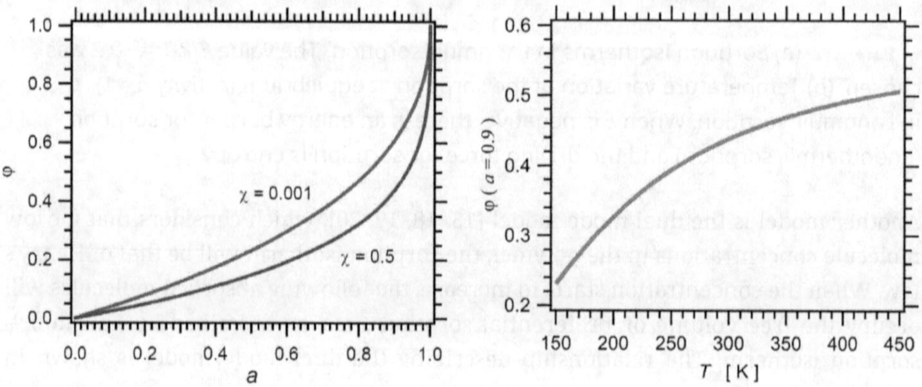

Figure 1.3 (Left) Sorption isotherms in the Flory model (Equation 1.8). (Right) Temperature variation of the sorption in the Flory model. The chosen value of the interaction parameter is $\chi = 0.5 \times 300$ K

Since $\chi > 0$, this always corresponds to endothermic sorption. The exponential argument contains an enthalpic term related to χ and an entropic term, which gives $1 + \chi$. Molecules have to overcome an energy barrier in order to adsorb, and the driving force for adsorption is entropy. Thus, it is expected that sorption will always increase as temperature increases (see Figure 1.3b).

In order to describe systems with strong interactions between the penetrant and the polymer, other models were developed, which either combine the previous ones or have a different approach. Among them, the Brunauer, Emmett, and Teller model [22] (BET II, see Figure 1.4) assumes there exists preferential sorption sites like in the Langmuir mode, but more than one molecule may be sorbed in each site. The first sorbed molecule in each site (energy E_1) is differentiated from additional layers (energy E_2, see Figure 1.4). Sorption isotherms are described by the Guggenheim, Anderson, and De Boer (GAB) equation [23–27], which is derived by writing the equilibrium conditions of successive layers:

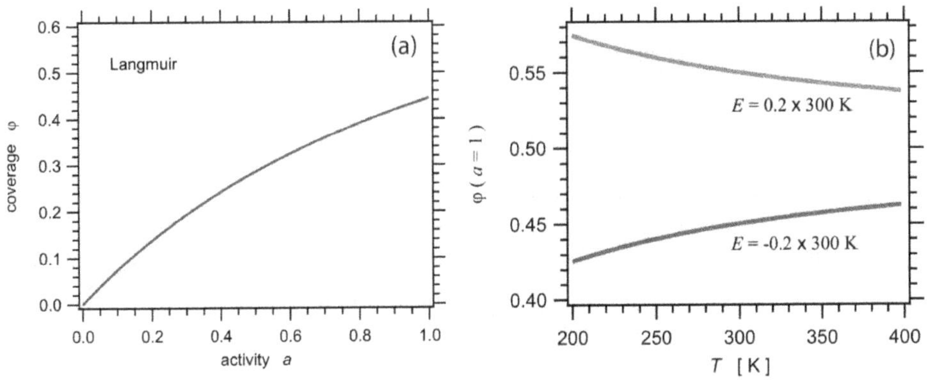

Figure 1.2 (a) Sorption isotherms in Langmuir sorption. The value $E/k_BT = -0.2$ was chosen. (b) Temperature variation of the sorption at equilibrium (activity $a = 1$) in Langmuir sorption. When E is negative, there is an energy barrier for sorption (endothermic sorption) and the driving force for sorption is entropy

Another model is the dual mode model [15, 18, 19, 20], which considers that for low molecule concentrations in the polymer, the sorption isotherm will be that of Henry's law. When the concentration starts to increase, the following absorbed molecules will occupy the free volume or preferential sorption sites, as described by Langmuir's sorption isotherm. The relationship describing the dual mode model is shown in Equation 1.6:

$$C = C_D + C_H = K_D a + C'_H \frac{ba}{1 + ba} \tag{1.6}$$

Where:

C_D and C_H = Henry and Langmuir molecule concentration contributions

C'_H and b = Langmuir's capacity and affinity constant respectively

The Flory–Huggins sorption mode [2–9] is used to describe non-polar polymer–solvent mixtures. The affinity between the penetrant and the polymer matrix is described by an interaction parameter χ, a dimensionless measurement of the differences in interaction energies between species i and j in a mixture, which can be estimated with:

$$\chi = \frac{(\delta_i - \delta_j)^2}{RT} \sqrt{v_i v_j} \tag{1.7}$$

Where:

δ_i = Hildebrand solubility parameter

v_i = molar volume of species i (solvent or monomer)

Using the expression of the free energy of mixing between a very long polymer chain and a solvent, sorption isotherms are described by [2–9]:

$$\ln a = \ln \phi + (1 - \phi) + \chi(1 - \phi)^2 \tag{1.8}$$

Where:

ϕ = solvent volume fraction in the polymer

In the limit of small sorption, Equation 1.8 reduces to $\phi = a \exp(-(1 + \chi))$, which corresponds to Henry's law. A typical isotherm is shown in Figure 1.3a.

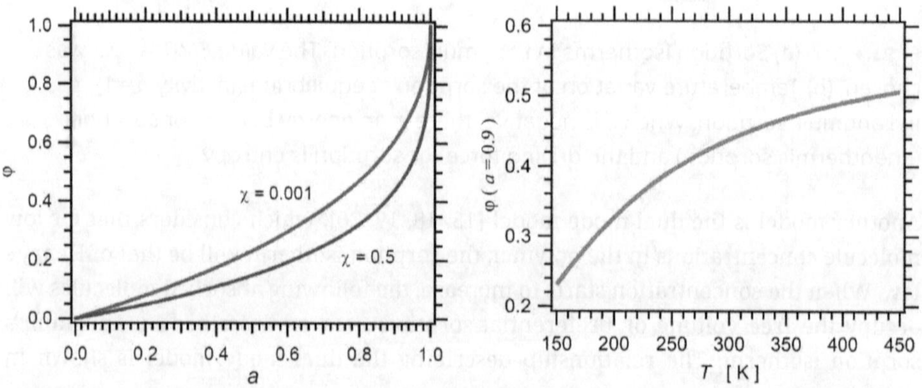

Figure 1.3 (Left) Sorption isotherms in the Flory model (Equation 1.8). (Right) Temperature variation of the sorption in the Flory model. The chosen value of the interaction parameter is $\chi = 0.5 \times 300$ K

Since $\chi > 0$, this always corresponds to endothermic sorption. The exponential argument contains an enthalpic term related to χ and an entropic term, which gives $1 + \chi$. Molecules have to overcome an energy barrier in order to adsorb, and the driving force for adsorption is entropy. Thus, it is expected that sorption will always increase as temperature increases (see Figure 1.3b).

In order to describe systems with strong interactions between the penetrant and the polymer, other models were developed, which either combine the previous ones or have a different approach. Among them, the Brunauer, Emmett, and Teller model [22] (BET II, see Figure 1.4) assumes there exists preferential sorption sites like in the Langmuir mode, but more than one molecule may be sorbed in each site. The first sorbed molecule in each site (energy E_1) is differentiated from additional layers (energy E_2, see Figure 1.4). Sorption isotherms are described by the Guggenheim, Anderson, and De Boer (GAB) equation [23–27], which is derived by writing the equilibrium conditions of successive layers:

$$M(a) = \frac{M_m ACa}{(1 - Aa)(1 + Aa(C - 1))} \tag{1.9}$$

Where:

$C = \exp((E_1 - E_2)/k_B T)$

$A = \alpha \exp(E_2/k_B T)$

$M(a)$ = solvent uptake at activity a

M_m = solvent content corresponding to saturation of all primary absorption sites by one solvent molecule (monolayer)

C =Guggenheim constant (and represents the difference in the adsorption energy for the first layer and the other successive layers)

A = a factor correcting the properties of the multilayer molecules with respect to the bulk liquid

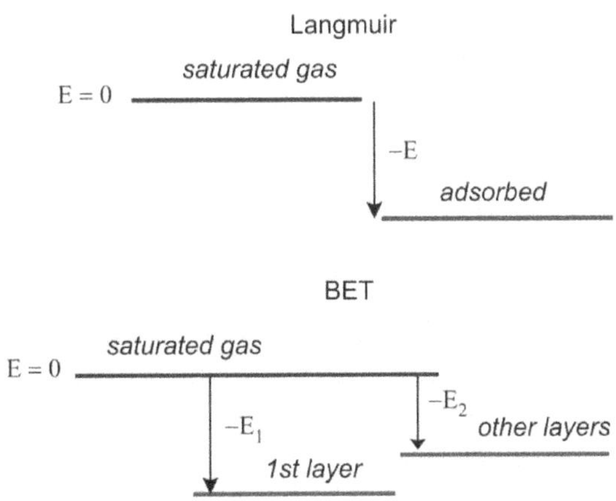

Figure 1.4 Adsorption energies in the Langmuir (monolayer) and BET (multilayer) cases. Reference level for energies is the saturated gas. In the schematized case, $E_2 > 0$ (E_2 is analogous to the heat of liquefaction) and $\Delta E > 0$)

In the limit of low activity ($a \lll 1$), Equation 1.9 reverts to the Langmuir case (because only one molecule per site is sorbed in the limit of small sorption):

$$\phi = \frac{ca}{1 + ac} \tag{1.10}$$

Another model that can describe the sorption of a molecule that interacts with the polymer is the ENSIC (engaged species induced clustering) model [28–32]. This model takes into account that molecules will be absorbed because they have strong affinities with the polymer and/or with the same type of molecules already present in the polymer. The volume fraction ϕ of solvent in the polymer is then defined by Equation 1.11:

$$\phi = \frac{e^{(k_S - k_P)a} - 1}{(k_S - k_P)/k_P} \tag{1.11}$$

Where:

k_P and k_S = affinity of an absorbed molecule with the polymer or another sorbed molecule respectively

In all the preceding sorption models, except for Henry's law, solvent molecules in multilayers on a sorption site are considered to be absobed into clusters. Empirical relationships allowing calculation of their size (i. e., mean cluster size, MCS) have been used for the GAB, ENSIC and Flory–Huggins isotherms.

In the case of the GAB model, the parameters obtained by this isotherm can also be used to estimate the MCS by Equation 1.10 [33, 34]:

$$\text{MCS} = -(1-\phi)\left[\frac{\phi}{M_m C}(-2 + 2Aa - 2ACa + C) - 1\right] \tag{1.12}$$

For the ENSIC model, the MCS can be estimated by Equation 1.13 [28, 29]:

$$\text{MCS} = \frac{(1-\phi)(k_2\phi + 1)\ln(k_2\phi + 1)}{k_2\phi} \tag{1.13}$$

Where:

$k_2 = \frac{k_S - k_P}{k_P}$

Finally, the MCS can also be calculated from the Flory–Huggins model according to Equation 1.14 [35]:

$$\text{MCS} = \frac{1}{1 - 2\chi\phi + (1-\phi)\left(\frac{\partial\chi}{\partial \ln\phi}\right)_{T,P}} \tag{1.14}$$

As we have seen, most of the proposed models use empirical parameters to explain the sorption regimes of molecules in polymers. In the following section, we derive a rigorous physical approach to describe the sorption of solvent molecules in a polymer with preferential absorption sites. The situation which is described is analogous to the BET sorption model, as more than one molecule can be sorbed in each site. However, in contrast to previous models, which are empirical or semi-empirical, the proposed approach is purely physical. This means that the proposed model yields quantitative parameters with physical meaning.

1.3 Model with Solvent with Adsorption Sites (H-Bonds)

The treatment described here is analogous to that of Cassie and Hill [36–38] (although the latter is not fully correct), and should be exactly analogous to GAB sorption model. The objectives are twofold. First, we want to describe the sorption isotherm ϕ (water content) as a function of water activity a_w (or equivalently the water partial pressure). Second, the way in which water molecules distribute in the material is described. This is intended to provide a quantitative way of characterizing the often ill-defined notion of water clusters and the distribution of cluster sizes.

A PA sample contains N polar groups (amide groups) per unit volume with which water molecules can establish hydrogen bonds, and these amide groups may thus be considered as absorption sites. Amide groups provide both a proton donor moiety (–NH) and a proton acceptor (>C=O). A water molecule can thus bond the amide groups in two different ways. It is generally considered that bonding to the proton acceptor (>C=O) moiety is favored [39]. In a first step, we shall not consider this possibility of double hydrogen bonding and assume that the material has N absorption sites per unit volume.

Water molecules absorbed in the material can be in two distinct states (see also [40]), either bonded to an absorption site (sorption energy ε_1), or bonded to another water molecule (sorption energy ε_2). Only the statistical distribution of water molecules is discussed here. The observation timescale must be long compared to the lifetime of a water molecule in a given state or to the timescale of exchange between both types of sites. This exchange dynamics depends on energy barriers that must be crossed as a molecule undergoes exchange between two distinct sites. There is no "free" water molecule, meaning that the lifetime of a bonded state is assumed to be very long with respect to the "time of flight" inbetween two distinct sites. Also, there is no term associated with swelling of the sample by water. Only polar interactions are implemented.

Orders of magnitude of sorption energies are as follows. The enthalpy of vaporization of water is 44 kJ/mol. The sorption enthalpy should thus be of the same order, while the difference $\Delta\varepsilon = \varepsilon_1 - \varepsilon_2$ should be comparatively small.

Assuming that a water molecule in the material can form a stronger bond with an amide group than with another water molecule is equivalent to saying that $\Delta\varepsilon = \varepsilon_1 - \varepsilon_2 < 0$ (i. e. the water–amide bond is more stable than the water–water bond, as illustrated in Figure 1.5).

On the other hand, it may be assumed that the material does not dissolve in water when immersed in the liquid, which means that water molecules have an energy barrier to pass to enter the material, or equivalently, that $\varepsilon_2 > 0$, or, in other words, the enthalpy of sorption is not equal to (in fact, is smaller than) the enthalpy of liquefaction, which is the reference level for enthalpies or free enthalpies in the liquid state (see Figure 1.5).

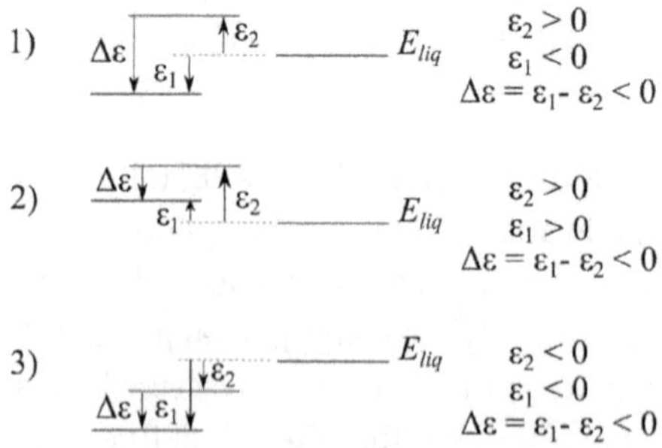

Figure 1.5 Scenarios for bonding of a water molecule with an amide group (ε_1) or another water molecule (ε_2). In case (3), an infinite quantity of water will be absorbed in liquid water, corresponding to dissolution of the polymer. More interesting cases are (1) and (2)

For a sample containing n water molecules per unit volume, the (macroscopic) configuration of the sample is described by the molar fraction of water $\Phi = n/N$ (referred to the number of absorption sites). Note that Φ can be larger than 1. The number n of water molecules is adjusted by the system according to the chemical potential of water, which is imposed through outer pressure. The other variable is the number m of occupied sites, which describes the way in which water molecules distribute in the material. A configuration in which n water molecules occupy exactly m sites over N available sites ($m < N$ and $m < n$) has a total bonding energy (or energy difference) $E(m, n) = m\varepsilon_1 + (n - m)\varepsilon_2 = n\varepsilon_2 + m\Delta\varepsilon$. The total number of such (microscopic) configurations is denoted $W(m, n)$ (or equivalently $W(m, \Phi)$), as seen in Figure 1.6.

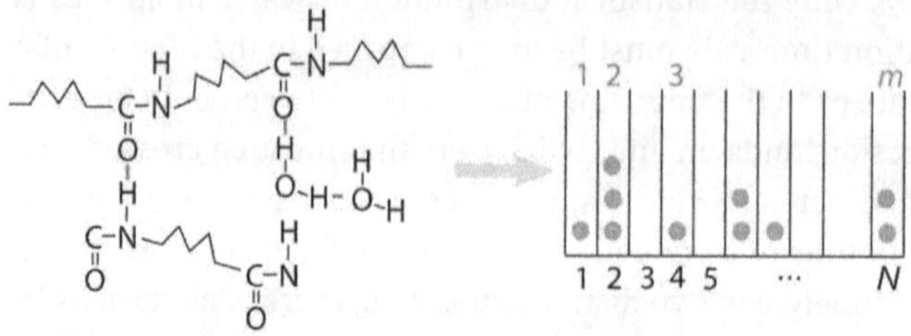

Figure 1.6 An example of a PA configuration. Each box represents one absorption site (a –CONH– group)

The partition function for a sample with a number n of water molecules (or equivalently a molar fraction of water $\Phi = n/N$) is:

$$Z(\Phi, \Delta\varepsilon) = \exp(-n\varepsilon_2) \sum_{m=1}^{N_0} W(m, n) \exp(-m\,\Delta\varepsilon) \tag{1.15}$$

Where:

$N_0 = \inf(N, n)$

Then the total free energy of the system, considered as a function of n (or equivalently Φ) and $\Delta\varepsilon$, is:

$$F(\Phi, \Delta\varepsilon) = -k_B T \ln Z(\Phi, \Delta\varepsilon) \tag{1.16}$$

1.3.1 Statistics of Distribution of Water Molecules

$W(m, n)$ in Equation 1.15 now has to be calculated. For $m \leq N_0$ for fixed m (or equivalently fixed Φ), let us first compute the number of ways to distribute n molecules in exactly m sub-ensembles (with at least one molecule in each sub-ensemble, $n \geq m$) as schematized in Figure 1.7.

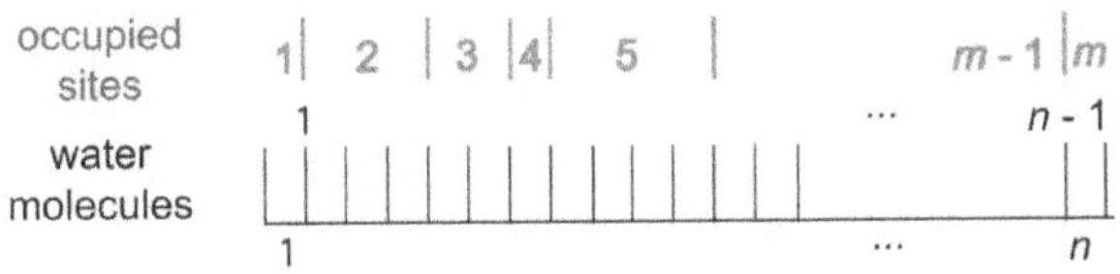

Figure 1.7 Distribution of n water molecules among exactly m absorption sites. Numbers in red label the successive occupied sites. Each site has a certain number of water molecules (a "water cluster") attached to it; for instance the second occupied site has three water molecules

For n molecules there are $n - 1$ "intervals" between molecules. Then $m - 1$ "separators" have to be placed among these $n - 1$ intervals, which gives a total number of configurations as follows:

$$X_m^n = C_{n-1}^{m-1} = \frac{(n-1)!}{(m-1)!(n-m)!} \approx \frac{n!}{m!(n-m)!} \tag{1.17}$$

This number has to be multiplied by the number of ways to select m sites among the list of N available sites:

$$C_N^m = \frac{N!}{m!(N-m)!} \tag{1.18}$$

giving $W(m, n)$ as:

$$W(m,n) = X_m^n C_N^m = \frac{n!}{m!(n-m)!} \frac{N!}{m!(N-m)!} \tag{1.19}$$

Using Stirling's approximation:

$$\ln W(m,n) = -(n-m)\ln\left(\frac{n-m}{n}\right) - m\ln\left(\frac{m}{n}\right) - (N-m)\ln\left(\frac{N-m}{N}\right) - m\ln\left(\frac{m}{N}\right) \tag{1.20}$$

Let us denote $\nu = \frac{m}{N}$ as the fraction of occupied absorption sites, and $u = \frac{m}{n} = \frac{\nu}{\Phi}$ (with $u < 1$). $\frac{1}{u} = \frac{\Phi}{\nu}$ is the average number of water molecules per occupied site (or equivalently the average size of "water clusters"). Then W may be expressed as a function of the variables Φ and ν as:

$$\ln W(\Phi, v) = Nf(v, \Phi) \tag{1.21}$$

with

$$f(v, \Phi) = -\Phi[(1 - v\Phi^{-1})\ln(1 - v\Phi^{-1}) + v\Phi^{-1}\ln v\Phi^{-1}] - (1 - v)\ln(1 - v) - v\ln v \tag{1.22}$$

Where:

$f(v, \Phi)$ = entropy per sorption site

v varies between 0 $\Phi_0 = \inf(1, \Phi)$. This is plotted in Figure 1.8. Then, the partition function $Z(\Phi, \Delta E)$ can be written (see Equation 1.15) as:

$$Z(\Phi, \Delta E) = \int_0^{\Phi_0} \exp[Nf(v, \Phi) - Nv\Delta\varepsilon - N\Phi\varepsilon]N\mathrm{d}v \tag{1.23}$$

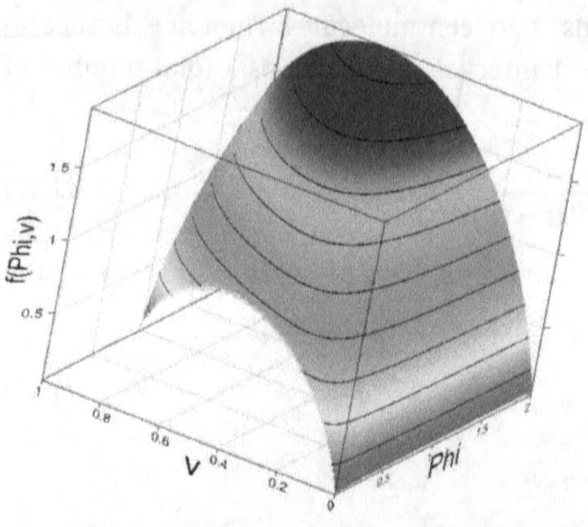

Figure 1.8 3D plot of $f(v, \Phi)$ according to Equation 1.22

1.3.2 Mean Field Approximation

In the limit $N \to \infty$, for a given value of Φ, only v values close to the value $\overline{v}(\Phi)$ corresponding to the maximum of $g(v, \Phi) = f(v, \Phi) + v\Delta\varepsilon + \Phi\varepsilon_2$ for each value of Φ contribute to the integral, and $Z(\Phi, \Delta E)$ may be approximated by:

$$\overline{Z}(\Phi, \Delta\varepsilon) = \sqrt{\frac{2\pi N}{|g''(\overline{v})|}} \exp\left[Nf(\Phi, \overline{v}) - N\overline{v}\Delta\varepsilon - N\Phi\varepsilon_2\right] \tag{1.24}$$

where symbols with an overbar denote mean-field quantities, and the prefactor corresponds to logarithmic corrections that are asymptotically negligible in the limit $N \to \infty$. Taking $F = -k_B T \ln Z$ with Z given by Equation 1.24, the mean field free energy per absorption site (amide group), as a function of Φ and in units of $k_B T$, is given by:

$$\overline{F}(\Phi, \Delta\varepsilon, \varepsilon_2) = -f\left(\Phi, \overline{v}\right) + \overline{v}\Delta\varepsilon + \Phi\varepsilon_2 \tag{1.25}$$

Where:

$f(\Phi, \overline{v})$ = entropy (given by Equation 1.22)

$\overline{v}\,\Delta\varepsilon + \Phi\varepsilon_2$ = interaction energy

$\overline{v}$ is given as a function of Φ by:

$$\left(\frac{\partial f}{\partial v}\right)_{\Phi} - \Delta\varepsilon = 0 \tag{1.26}$$

which gives:

$$\ln\left(\frac{1-\overline{v}\Phi^{-1}}{\overline{v}\Phi^{-1}}\right) + \ln\left(\frac{1-\overline{v}}{\overline{v}}\right) - \Delta\varepsilon = 0 \tag{1.27}$$

or equivalently:

$$\Phi = \frac{\overline{v} + \overline{v^2}(e^{\Delta\varepsilon} - 1)}{1-\overline{v}} \tag{1.28}$$

Solving for $\overline{v}$ and writing $\kappa = \exp(\Delta\varepsilon - 1)$ (or $\Delta\varepsilon = \ln(\kappa + 1)$), this gives:

$$\overline{v} = \frac{-(1+\Phi) + \sqrt{(1+\Phi)^2 + 4\kappa\Phi}}{2\kappa} \tag{1.29}$$

Note that in the limit $\Phi \lll 1$, $\overline{v} \approx \Phi$. For $\Delta\varepsilon = 0$ (i. e. $\kappa = 0$), Equation 1.28 gives:

$$\overline{v_0}(\Phi) = \frac{\Phi}{1+\Phi} \tag{1.30}$$

An expansion of $\overline{v}\,(\Phi)$ (at a given fixed value of Φ) in the vicinity of $\kappa = 0$ is:

$$\overline{v} \approx \left(\frac{\Phi}{1+\Phi}\right)\left[1 - \frac{\kappa\Phi}{(1+\Phi)^2} + \frac{\kappa^2\Phi^2}{(1+\Phi)^4}\right] \tag{1.31}$$

We can now approximate the average "cluster size" as the following equation (represented in Figure 1.9):

$$\frac{\Phi}{\overline{v}} \approx (1+\Phi)\left[1 + \frac{\kappa\Phi}{(1+\Phi)^2} + o(\Phi^3)\right] \tag{1.32}$$

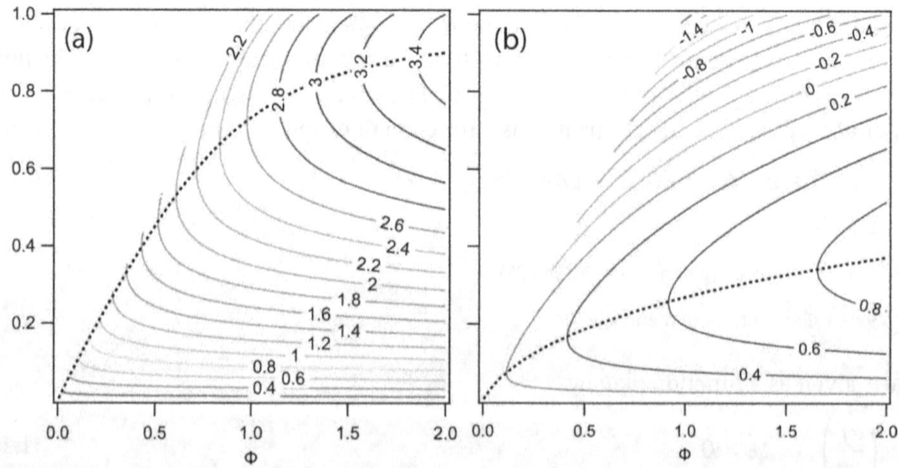

Figure 1.9 Contour plot of $f(v, \Phi)$ (according to Equation 1.22). The dashed curve is the function $\bar{v}(\Phi)$ as defined in Equation 1.28. (a) $\Delta\varepsilon = +2.0$ (weak aggregation); (b) $\Delta\varepsilon = -2.0$ (stronger aggregation) in k_BT units

1.3.3 Sorption Isotherms

We must now determine the amount Φ of absorbed water for a given water activity outside the sample. A fixed quantity of polymer (number of adsorption sites N corresponding to a unit volume of dry polymer) is put in the presence of solvent (water) in the vapor phase at partial vapor pressure p (or activity a) or chemical potential $\mu = \mu_0 + \ln a$ (in units of k_BT) (Figure 1.10) with $a = p/p_{sat}$ the water activity (p_{sat} is the saturated vapor pressure). μ_0 refers to the saturated atmosphere (or equivalently to the liquid phase).

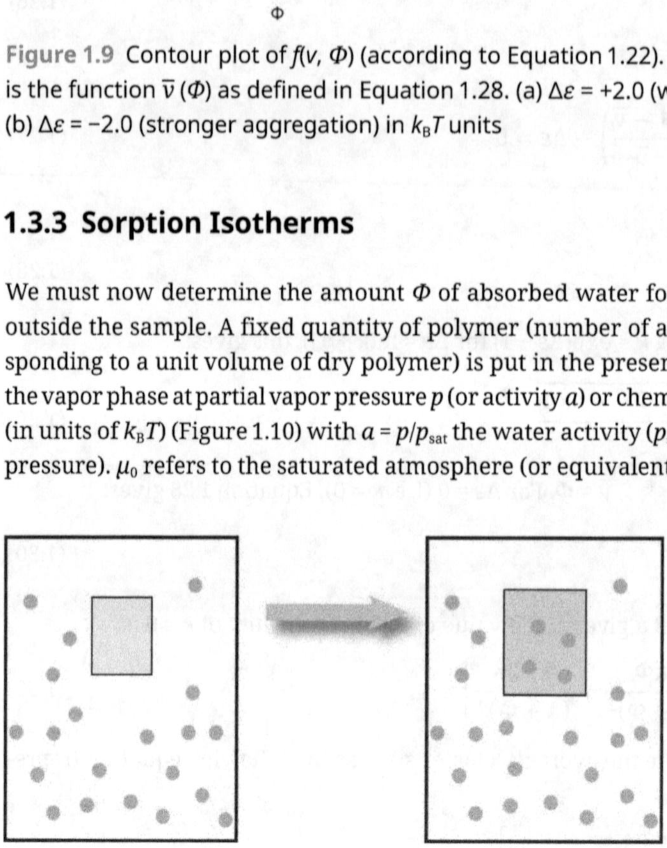

Figure 1.10 Equilibrium of the system in the presence of water vapor

For a given value of $\Delta\varepsilon$ and ε_2, the polymer absorbs n molecules of water (fraction $\Phi = n/N$). The equilibrium water content Φ_{eq} is determined by the condition:

$$\mu = \left(\frac{\partial F_{tot}}{\partial n}\right)_N = \mu_{vap} = \mu_0 + k_BT \ln \frac{p}{p_{sat}} \tag{1.33}$$

The total mean field free energy $F_{tot} = N\overline{F}(\Phi, \Delta\varepsilon)$ is given by Equation 1.25, giving an **explicit expression for the sorption isotherm**, depending on parameters κ and ε_2:

$$\ln\frac{p}{p_{sat}} = \ln a = \left(\frac{\partial F_{tot}}{\partial n}\right)_N = \left(\frac{d\overline{F}}{d\Phi}\right)_N = -\left(\frac{\partial f}{\partial \Phi}\right)_{\overline{\nu}} - \frac{\partial \overline{\nu}}{\partial \Phi}\left(\left(\frac{\partial f}{\partial \overline{\nu}}\right)_{\Phi} - \Delta\varepsilon\right) + \varepsilon_2 \quad (1.34)$$

with f given in Equation 1.22 and $\overline{\nu}$ in Equation 1.29.

For small κ values, an expansion up to second order in κ is:

$$\ln a = \ln\left(\frac{\Phi}{1+\Phi}\right) + \varepsilon_2 + \frac{\kappa}{(1+\Phi)^2} - \frac{3\kappa^2(2\Phi - 1)}{2(1+\Phi)^4} \quad (1.35)$$

The resulting sorption isotherms as a function of the chemical activity a are given in Figure 1.11a, and the evolution of the MCS as a function of the sorbed solvent fraction Φ are given in Figure 1.11b.

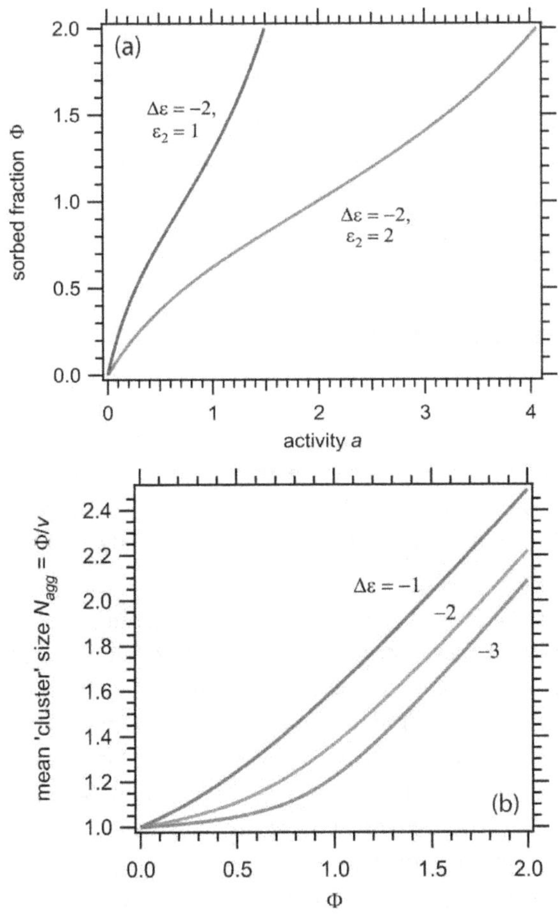

Figure 1.11
(a) Sorption isotherms according to Equation 1.33 and Equation 1.34, and (b) mean cluster size given by $\Phi/\overline{\nu}$ as a function of Φ from the expression for $\overline{\nu}$ given by Equation 1.29

1.4 Comparison with Experimental Results

In order to assess the physical relevance of the proposed sorption model, selected polar solvent–polar polymer systems were analyzed. Two polymer matrices were considered: polyamides and polyacrylic acid. Polyamides are semi-crystalline polymers obtained by a polycondensation reaction, and have been widely used in commodity and technical applications since their invention in the 1930s (clothing fibers, technical carbon fibers, hoses, automotive parts, etc.) [39]. In contrast, polyacrylic acid is a completely amorphous polymer obtained by radical polymerization that is mostly used in the cosmetics industry as an emulsification agent [41, 42].

The first two systems we discuss are polyamide–water (Section 1.4.1) and PAA–water (Section 1.4.2), since water is one of the most common solvents towards which polymers have to act as a barrier. The third system considered is polyamide–alcohol (Section 1.4.3), because of the use of polyamides as hoses in gasoline pumps, as well as for conveying gasoline inside cars. With the introduction of ethanol into gasoline, which is mainly composed of hydrocarbons, the sorption and diffusion of gasoline in these materials changes due to the interactions between ethanol and polyamides [43, 44].

These three comparisons demonstrate the behavior of two completely different polymers in contact with two different solvents, showing the versatility and applicability of the model for different polymer–solvent systems.

1.4.1 Sorption of Water into Polyamides

One of the most extensive studies of sorption of water into polyamide 6 (PA6) was by Skirrow and Young [11], with the original data shown in Figure 1.12a. Selected data are re-plotted in Figure 1.12b in terms of the molar fraction Φ of sorbed water (as defined above) as a function of the activity $a(T) = p/p_{sat}(T)$. Other data in the literature is that of Sabard for PA6 [45, 46] and Preda for PA6,6 and PA6,10 [47]. These data are also analyzed below.

Water, or any other solvent, supposedly does not enter crystallites, meaning that its sorption occurs solely on the amorphous phase [48–50]. The molar fraction of water in the amorphous phase should then be corrected upwards compared to the overall fraction to take the (non-permeable) crystalline fraction into account [43]. The crystalline ratio of the samples is not reported in the study by Skirrow et al. In this work it was thus considered that the crystalline fraction was of 36% for PA6 [39], 38% for PA6,6 [43, 44, 47], and 23% for PA6,10 [47]. Moreover, in these calculations it has been assumed that solvent sorption occurs homogeneously in the amorphous phase. Indeed, as the amorphous phase possesses a gradient of molecular mobility (exemplified by mobile and rigid amorphous fractions, MAF and RAF [48–50]), it might be possible that solvent sorption would be affected by this molecular mobility gradient, i. e., water would be absorbed easily by the MAF, while the RAF would be less accessible for this molecule. In this work the water sorption data was recalculated so as to reflect that water sorption occurs homogeneously in the amorphous phase.

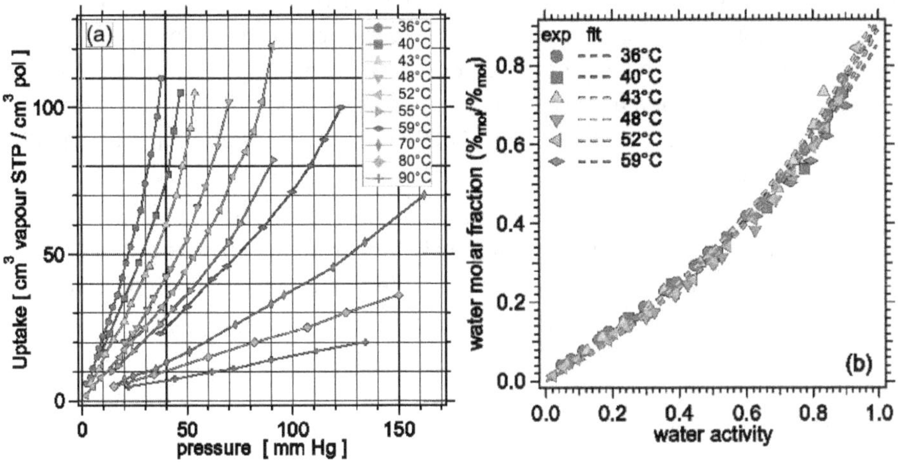

Figure 1.12 (a) Data for the sorption of water in PA6 from ref. [11]; (b) Selected data for the sorption of water in PA6, plotted differently than in ref. [11]. Symbols are measured data; curves are fits with the model introduced in Section 1.3

Figure 1.12b shows the recalculated data from Skirrow. As expected, all curves measured at different temperatures nearly collapse onto each other, as the strong variation of the saturating vapor pressure as a function of temperature is eliminated in this representation. Recalculated data from the work of Sabard [45, 46] and Preda [47] are plotted in Figure 1.13.

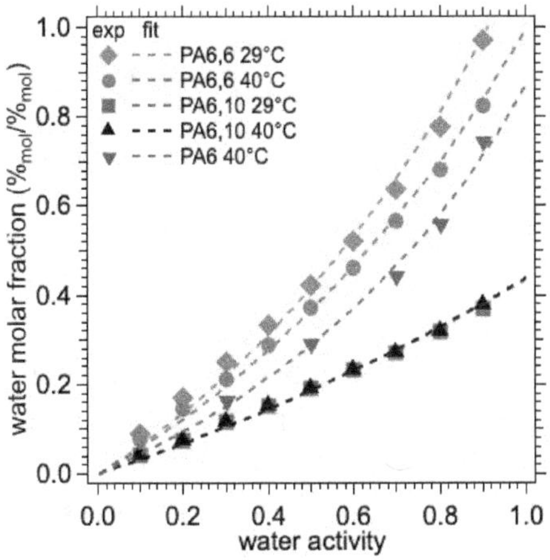

Figure 1.13
Data for the sorption of water in PA6, PA6,6, and PA6,10, reproduced from ref. [45, 46, 47]. Symbols are measured data; curves are fits with the model introduced in Section 1.3

The experimental data in Figure 1.12b and Figure 1.13 were fitted according to the proposed model in Equation 1.35. A very good agreement between the experimental results and each individual fit for each temperature and polyamide is observed. From these fits, the values of ε_2 and κ were obtained. From these data, the values of $\Delta\varepsilon$ and ε_1 were further calculated from $\Delta\varepsilon = \ln(\kappa + 1)$ and $\Delta\varepsilon = \varepsilon_1 - \varepsilon_2$. These values are listed in Table 1.1 and are compared to calculated and/or measured ΔE energies for model amide–water systems. For such model systems, ΔE represents the gain in energy for a molecule of water to de-bond from another water molecule and bond to an amide group. These values should be compared to the values of ε_1 and ε_2 obtained by the fits. ΔE data for comparison was taken from Mitchell [51] and Dixon [52], and are also listed in Table 1.1.

Table 1.1 Calculated $\Delta\varepsilon$, ε_1, and ε_2 Energies (in kJ/mol) from the Proposed Model in Equation 1.35 for Water–PA6 Systems at Selected Temperatures, Compared to Water–Amide Interaction Energies ΔE (in kJ/mol)

System	Ref.	Temperature (°C)	ΔE	$\Delta\varepsilon$	ε_1	ε_2
Water–amide	Mitchell et al. [51]	—	2.3–4.2	—	—	—
	Dixon et al. [52]	—	2.1–3.5	—	—	—
	Skirrow [11]	—	0.4–4.6	—	—	—
Water–PA6	Skirrow [11]	36	—	-0.60	1.48	2.08
		40	—	-0.53	1.64	2.17
		43	—	-0.26	1.70	1.97
		48	—	-0.14	1.86	2.00
		52	—	-0.05	1.99	2.04
		59	—	-0.56	1.62	2.51
	Sabard [45, 46]	40	—	-0.29	1.62	1.91
Water–PA6,6	Preda [47]	29	—	-0.90	0.86	1.76
		40	—	-0.85	1.20	2.05
Water–PA6,10	Preda [47]	29	—	-0.54	2.71	3.25
		40	—	-0.55	2.80	3.35

It can be seen in Table 1.1 that the calculated values of ε_1 and ε_2 are in very good agreement with the ΔE values for all selected temperatures and polyamides. Furthermore, from the κ values obtained through the fits from Equation 1.35, the MCS evolution for water sorption was calculated according to Equation 1.32. These results are shown in Figure 1.14a for the data from Skirrow [11], and in Figure 1.14b for the data from Sabard [45, 46] and Preda [47].

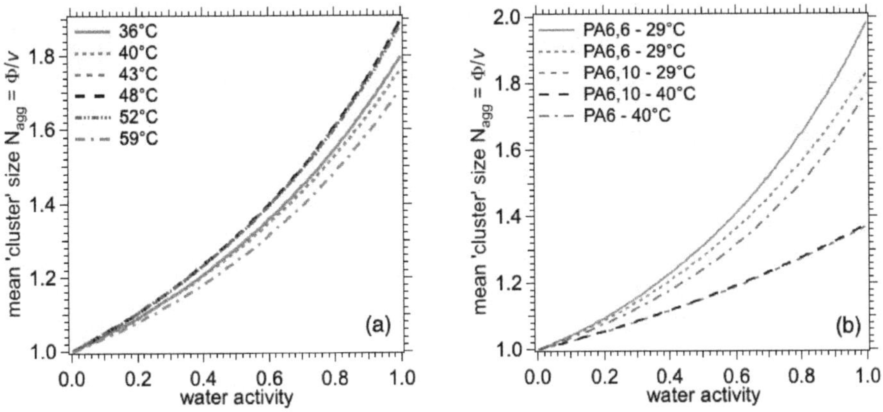

Figure 1.14 Evolution of the mean "cluster" size for water sorption as a function of its activity, as calculated from Equation 1.32 for: (a) data from ref. [11] in PA6 for selected temperatures; and (b) in PA6,6, PA6,10, and PA6 from ref. [45–47]

The data plotted in Figure 1.14a and Figure 1.14b, specifically the MCS values obtained at an activity of $a \geq 0.9$, were then compared to data available in the literature (Table 1.2).

Table 1.2 MCS Values from the Model Developed in Equation 1.35 for Water–Polyamide Systems, Compared to Those from Other Models

Model	Ref.	Polyamide	Temperature (°C)	MCS
GAB	Starkweather [53, 54]	PA6,6	25	3.0
	Lim [55]	PA6,6	23	3.0
	Espuche [56]	PA6	22	1.6
	Sabard [45, 46]	PA6	40	2.3
	Preda [47]	PA6,6	29	2.0
			35	2.0
			40	1.7
		PA6,10	29	1.3
			40	1.3
			51	1.3

Table 1.2 MCS Values from the Model Developed in Equation 1.35 for Water–Polyamide Systems, Compared to Those from Other Models *(continued)*

Model	Ref.	Polyamide	Temperature (°C)	MCS
ENSIC	Sabard [45, 46]	PA6	40	1.6
Flory–Huggins	Sabard [45, 46]	PA6	40	2.3
Sotta–Ríos	Skirrow [11]	PA6	36	1.8
			40	1.8
			43	1.9
			48	1.9
			52	1.9
			59	1.7
	Sabard [45, 46]	PA6	40	1.8
	Preda [47]	PA6,6	29	2.0
			40	1.8
		PA6,10	29	1.4
			40	1.4

Table 1.2 shows that the MCS values calculated by the proposed model are in good agreement with those found in the literature. These results show the pertinence of the model.

1.4.2 Sorption of Water into Polyacrylic Acid

The same approach was undertaken for water–polyacrylic acid systems. Polyacrylic acid (PAA) is a completely amorphous polymer obtained from a radical polymerization reaction, in contrast to polyamides which are obtained by polycondensation. The original water sorption data for this polymer, obtained by Thijs [42] and Chang [41], are shown in Figure 1.15a. These data were re-plotted in Figure 1.15b, as in the previous case for PA6 (with the water molar fraction Φ as a function of the activity).

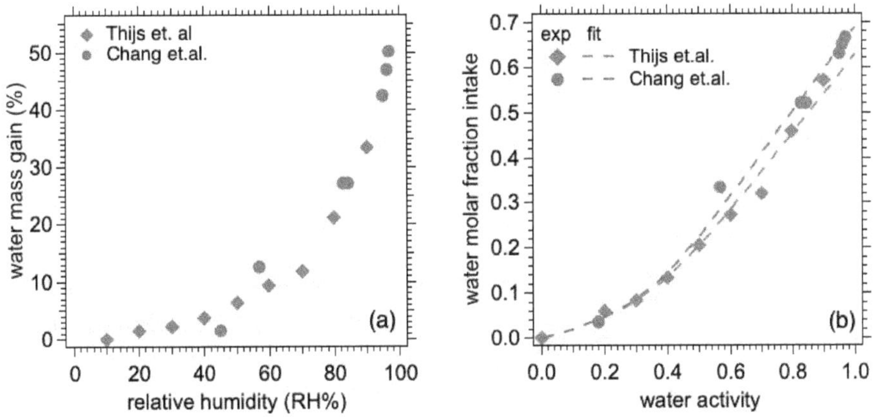

Figure 1.15 (a) Water mass intake percentage in PAA at 30 °C as a function of relative humidity. The data was taken from refs. [42, 41]; (b) The ame data plotted differently. Symbols are measured data, curves are fits with the model introduced in Section 1.3

In the case of water–PAA sorption, the experimental data in Figure 1.15b was fitted according to Equation 1.35. Again, a very good agreement between the experimental results from ref. [42, 41] and each individual fit is observed. The values of $\Delta\varepsilon$, ε_1, and ε_2 were obtained from such fits and are listed in Table 1.3. Such values are compared to ΔE energies for model water–carbonyl systems, in which ΔE represents this time the gain in energy for a molecule of water to de-bond from another water molecule and bond to a carboxylic group. These values, also listed in Table 1.3, are compared to ε_1 and ε_2.

Table 1.3 Calculated $\Delta\varepsilon$, ε_1, and ε_2 Energies (in kJ/mol) from the Model Developed in Equation 1.35 for Water–PAA Systems, Compared to Water-Carbonyl Interaction Energies ΔE (in kJ/mol)

Ref.	ΔE	$\Delta\varepsilon$	ε_1	ε_2	MCS
Morokuma [57]	1-1.42	—	—	—	—
Thijs [42]	—	-2.37	1.05	3.42	1.3
Chang [41]	—	-2.44	0.84	3.28	1.3

It is seen in Table 1.3 that again, the calculated values of ε_1 and ε_2 from Equation 1.35 for water–PAA systems [41, 42] are in very good agreement with the interaction energies ΔE from [57]. Then, from the κ values also obtained from Equation 1.35, the change in MCS with water sorption in PAA was calculated, and is plotted in Figure 1.16. It is seen that the evolution of the plots is similar for both datasets.

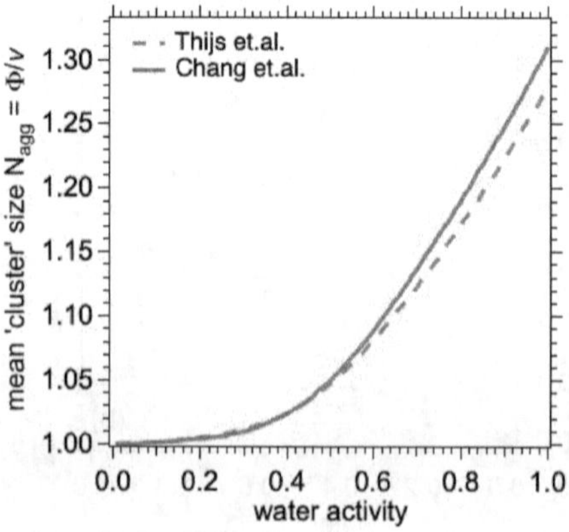

Figure 1.16 Evolution of the mean "cluster" size for water sorption in PAA as a function of its activity, as calculated from Equation 1.32 for data from refs. [41, 42]

1.4.3 Sorption of Alcohols into Polyamides

Skirrow and Young also studied the sorption isotherms of methanol in PA6 at several temperatures [11]. The original data are shown in Figure 1.17a, and selected data are re-plotted in Figure 1.17b in terms of the molar fraction Φ of sorbed methanol (as defined above) as a function of the activity $a(T) = p/p_{sat}(T)$. As in the case of water, the methanol sorption data was recalculated with regard to the crystalline ratio of PA6, to reflect that water sorption occurs homogeneously in the amorphous phase. As was observed for the sorption of water, data for methanol measured at different temperatures nearly collapse onto each other, as the strong variation of the saturating vapor pressure as a function of temperature is eliminated in this representation.

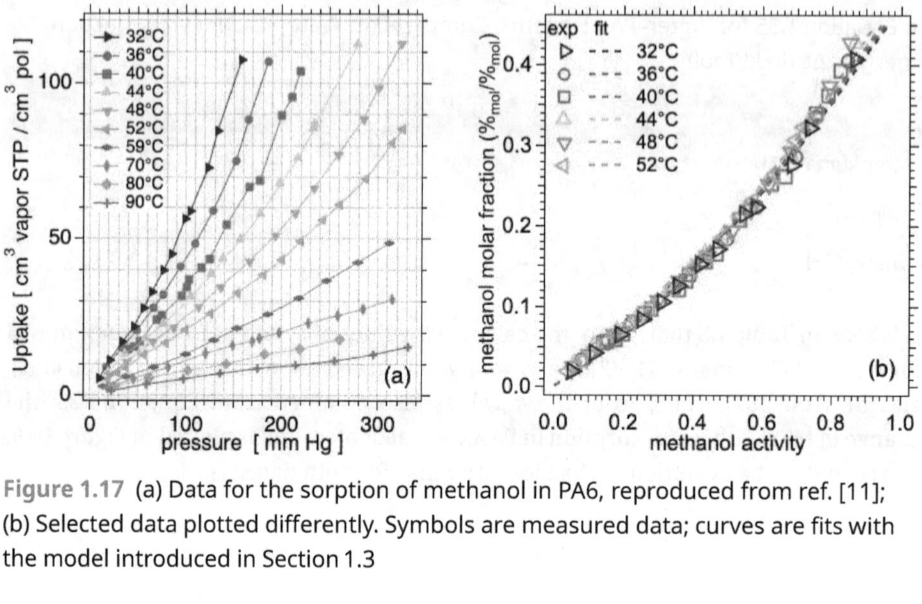

Figure 1.17 (a) Data for the sorption of methanol in PA6, reproduced from ref. [11]; (b) Selected data plotted differently. Symbols are measured data; curves are fits with the model introduced in Section 1.3

As was done previously, experimental data in Figure 1.17b were fitted according to Equation 1.35. A very good agreement between the experimental results and each individual fit for each temperature is observed. From these fits, the values of $\Delta\varepsilon$, ε_1, and ε_2 were obtained. These values are listed in Table 1.4 and are compared to calculated and/or measured ΔE energies for model methanol–polyamide systems. Again, these values should be compared to the values of ε_1 and ε_2 obtained by the fits. ΔE data for comparison was taken from Nagaraju [58], and are also listed in Table 1.4.

Table 1.4 Calculated $\Delta\varepsilon$, ε_1, and ε_2 Energies (in kJ/mol) from the Model Developed in Equation 1.35 for Methanol–PA6 Systems at Selected Temperatures, Compared to Alcohol–Carbonyl Interaction Energy Values ΔE (in kJ/mol)

Ref.	Temperature (°C)	ΔE	$\Delta\varepsilon$	ε_1	ε_2	MCS
Nagaraju [58]	—	2-5	—	—	—	—
Skirrow [11]	—	0.6–3.8	—	—	—	—
Skirrow [11]	32	—	-1.40	2.10	3.50	1.3
	36	—	-1.36	2.14	3.50	1.3
	40	—	-1.52	2.06	3.58	1.3
	44	—	-1.22	2.26	3.48	1.3
	48	—	-1.45	2.13	3.58	1.3
	52	—	-1.05	2.48	3.51	1.4

It can be seen in Table 1.4 that the calculated values of ε_1 and ε_2 are in very good agreement with those of ΔE from the literature. Furthermore, from the κ values obtained through the fits from Equation 1.35, the MCS evolution for water sorption was calculated according to Equation 1.32. These results are shown in Figure 1.18, with the values for $a \geq 0.9$ listed in Table 1.4.

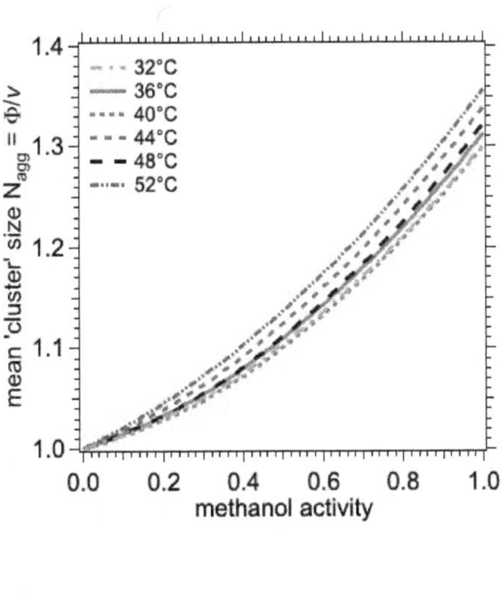

Figure 1.18
Evolution of the mean "cluster" size for methanol sorption in PA6 as a function of its activity, as calculated from Equation 1.32 for data from ref. [11]

Further available data in the literature is that of Sabard for the sorption of ethanol in PA6 [45, 46] and Preda for the sorption of ethanol in PA6,6 and PA6,10 [47]. These data were also analyzed herein, and are plotted in Figure 1.19a.

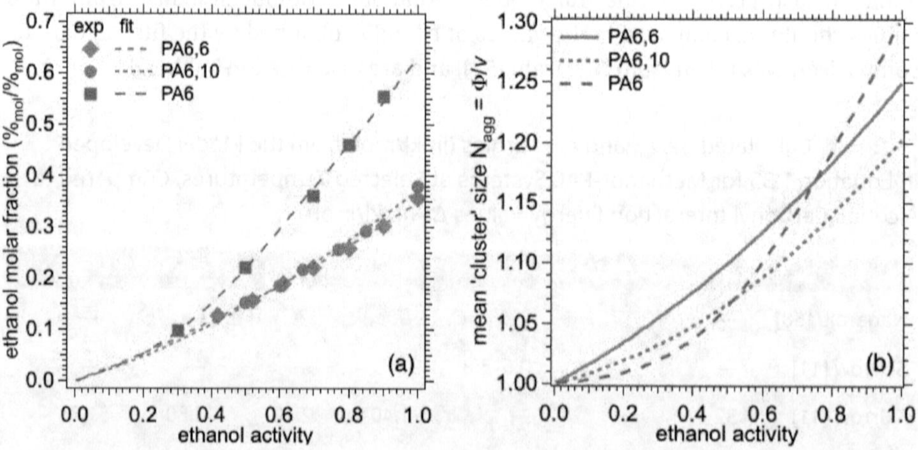

Figure 1.19 (a) Data for the sorption of ethanol in PA6, PA6,6 and PA6,10 reproduced from ref. [45–47]. Symbols are measured data; curves are fits with the model introduced in Section 1.3; (b) Evolution of the mean "cluster" size for ethanol sorption in PA6, PA6,6, and PA6,10 as a function of its activity, as calculated from Equation 1.32 for this same data

Again, the same approach was undertaken with this data. Curves were fitted according to Equation 1.35. The obtained values of $\Delta\varepsilon$, ε_1, and ε_2 are listed in Table 1.5 and are compared with data from Nagaraju [58] and Skirrow [11].

Table 1.5 Calculated $\Delta\varepsilon$, ε_1, and ε_2 Energies (in kJ/mol) from Equation 1.35 for Ethanol–Polyamide Systems at 40 °C, Compared to Alcohol–Carbonyl Interaction Energies ΔE (in kJ/mol)

Ref.	Polymer	ΔE	$\Delta\varepsilon$	ε_1	ε_2
Nagaraju [58]	—	2–5	—	—	—
Skirrow [11]	—	1.7–3.9	—	—	—
Sabard [45, 46]	PA6	—	-1.01	3.02	4.03
Preda [47]	PA6,6	—	-1.64	2.52	4.16
	PA6,10	—	-2.27	1.16	3.44

Table 1.5 shows that the calculated values of ε_1 and ε_2 are in very good agreement with those of ΔE from the literature. Moreover, from the κ values obtained through the fits from Equation 1.35, the MCS evolution for ethanol sorption was calculated according to Equation 1.32. These results are shown in Figure 1.19b, with the values for $a \geq 0.9$ listed in Table 1.6. These values are compared to those obtained by the GAB and ENSIC models, also from [45–47].

Table 1.6 Calculated MCS Values from Equation 1.32 for Ethanol–Polyamide Isotherms at 40 °C, Compared to MCS Values Obtained for These Isotherms from Various Models

Model	Ref.	Polyamide	MCS
GAB	Sabard [45, 46]	PA6	1.7
	Preda [47]	PA6,6	2.0
		PA6,10	2.0
ENSIC	Sabard [45, 46]	PA6	1.4
Sotta-Ríos	Sabard [45, 46]	PA6	1.3
	Preda [47]	PA6,6	1.3
		PA6,10	1.2

Table 1.6 shows that the MCS values calculated by the proposed model are again in good agreement with those listed in the literature. These results provide further confirmation of the performance of the model presented in this chapter.

1.5 Conclusion

This chapter has presented a new model for sorption of polar solvents in polar polymers. Compared to existing models that predict solvent intake and in some cases yield an empirical MCS formed by such solvents in polymers, this new model is able to give actual physical parameters, namely the gain in energy when a polar molecule interacts with a polar function in the polymer, and the MCS of this molecule for a given polymer calculated from this energy gain.

To validate this model, sorption at equilibrium of water as a function of its chemical activity in various polyamides and in polyacrylic acid were fitted and compared to values obtained by other models. The presented model accurately fitted the water sorption at equilibrium isotherms and yielded quantitative realistic values of the energy gain and the MCS for water when absorbed in these polymers. Moreover, the sorption of alcohols in polyamides was also modeled, yielding very satisfactory quantitative realistic values of the energy gain and the MCS for these solvents. The validation of the model by experimental results allows us to conclude that the proposed sorption model can be readily used for any polar polymer in contact with polar solvents.

References

[1] Langmuir, I. The constitution and fundamental properties of solids and liquids. Part I. Solids, *Journal of the American Chemical Society* 1916, 38, 2221–2295

[2] Huggins, M. Solutions of long chain compounds, *The Journal of Chemical Physics* 1941, 9, 440–440

[3] Flory, P.J. Thermodynamics of high polymer solutions, *The Journal of Chemical Physics* 1942, 10, 51–61

[4] Flory, P.J. *Principles of Polymer Chemistry*, Cornell University Press, 1953

[5] Flory, P.J.; Orwoll, R.; Vrij, A. Statistical thermodynamics of chain molecule liquids. I. An equation of state for normal paraffin hydrocarbons, *Journal of the American Chemical Society* 1964, 86, 3507–3514

[6] Flory, P.J.; Orwoll, R.A.; Vrij, A. Statistical thermodynamics of chain molecule liquids. II. Liquid mixtures of normal paraffin hydrocarbons, *Journal of the American Chemical Society* 1964, 86, 3515–3520

[7] Flory, P.J.; Abe, A. Thermodynamic properties of nonpolar mixtures of small molecules, *Journal of the American Chemical Society* 1964, 86, 3563–3565

[8] Flory, P.J. Statistical thermodynamics of liquid mixtures, *Journal of the American Chemical Society* 1965, 87, 1833–1838

[9] Abe, A.; Flory, P.J. The thermodynamic properties of mixtures of small, nonpolar molecules, *Journal of the American Chemical Society* 1965, 87, 1838–1846

[10] Crank, J. The Mathematics of Diffusion, Oxford University Press, 2004

[11] Skirrow, G.; Young, K.R. Sorption, diffusion and conduction in polyamide-penetrant systems: 1. Sorption phenomena, *Polymer* 1974, 15, 771–776

[12] Landau, L.D.; Lifshitz, E.M. *Statistical Physics, Vol. V: Theoretical Physics* (3rd edition), Mir Editions, 1967

[13] Mills, I.; Cvitas, T.; Homann, K.; Kallay, N.; Kuchitzu, K. *IUPAC Quantities, Units and Symbols in Physical Chemistry*, Blackwell Science, 1993

[14] Choudalakis, G.; Gotsis, A.D. Permeability of polymer/clay nanocomposites: A review, *European Polymer Journal* 2009, 45, 967–984

[15] Chandra, P.; Koros, W.J. Sorption of lower alcohols in poly (ethylene terephthalate), *Polymer* 2009, 50, 4241–4249

[16] Hansen, R.S.; Craig, R.P. The adsorption of aliphatic alcohols and acids from aqueous solutions by non-porous carbons, *The Journal of Physical Chemistry* 1954, 58, 211–215

[17] Dunicz, B.L. Surface area of activated charcoal by langmuir adsorption isotherm, *Journal of Chemical Education* 1961, 38, 357

[18] Paul, D.R. Gas sorption and transport in glassy polymers, *Berichte der Bunsengesellschaft für Physikalische Chemie* 1979, 83, 294–302

[19] Klopffer, M.H.; Flaconneche, B. Transport properties of gases in polymers: Bibliographic review, *Oil & Gas Science and Technology* 2001, 56, 223–244

[20] Guo, J.; Barbari, T.A. Unified dual mode description of small molecule sorption and desorption kinetics in a glassy polymer, *Macromolecules* 2009, 42, 5700–5708

[21] Hansen, C.M. *Hansen Solubility Parameters: A User's Handbook*, CRC Press, 2007

[22] Brunauer, S.; Emmett, P.H.; Teller, E. Adsorption of gases in multimolecular layers, *Journal of the American Chemical Society* 1938, 60, 309–319

[23] Anderson, R. B; Modifications of the Brunauer, Emmett and Teller equation, *Journal of the American Chemical Society* 1946, 68, 686–691

[24] de Boer, J. H. *The Dynamical Character of Adsorption*, Clarendon Press, 1953

[25] Guggenheim, E. A. *Applications of Statistical Mechanics*, Clarendon Press, 1966.

[26] Perrin, L.; Nguyen, Q. T.; Sacco, D.; Lochon, P. Experimental studies and modelling of sorption and diffusion of water and alcohols in cellulose acetate, *Polymer International* 1997, 42, 9–16

[27] Basu, S.; Shivhare, U. S.; Mujumdar, A. S. Models for sorption isotherms for foods: A review, *Drying Technology* 2006, 24, 917–930

[28] Favre, E.; Clément, R.; Nguyen, Q. T.; Schaetzel, P.; Néel, J. Sorption of organic solvents into dense silicone membranes. 2. Development of a new approach based on a clustering hypothesis for associated solvents, *Journal of the Chemical Society, Faraday Transactions* 1993, 89, 4347–4353

[29] Favre, E.; Nguyen, Q. T.; Clément, R.; Néel, J. The engaged species induced clustering (ENSIC) model: A unified mechanistic approach of sorption phenomena in polymers, *Journal of Membrane Science* 1996, 117, 227–236

[30] Jonquières, A.; Perrin, L.; Durand, A.; Arnold, S.; Lochon, P. Modelling of vapour sorption in polar materials: Comparison of Flory–Huggins and related models with the ENSIC mechanistic approach, *Journal of Membrane Science* 1998, 147, 59–71

[31] Elberaïchi, A.; Daro, A.; David, C. Water vapour transport in polyethylene oxide/polymethyl methacrylate blends, *European Polymer Journal* 1999, 35, 1217–1228

[32] Haidong, F. Prediction of infinite dilution solvent activity coefficient in rubbery polymer solutions using engaged species induced clustering (ENSIC) model, *Journal of Polymer Science Part B: Polymer Physics* 2006, 44, 1668–1675

[33] Zhang, Z.; Britt, I. J.; Tung, M. A. Water absorption in EVOH films and its influence on glass transition temperature, *Journal of Polymer Science Part B: Polymer Physics* 1999, 37, 691–699

[34] Dolmaire, N.; Espuche, E.; Méchin, F.; Pascault, J.-P. Water transport properties of thermoplastic polyurethane films, *Journal of Polymer Science Part B: Polymer Physics* 2004, 42, 473–492

[35] Marais, S.; Métayer, M.; Nguyen, T. Q.; Labbé, M.; Perrin, L.; Saiter, J. M. Permeametric and microgravimetric studies of sorption and diffusion of water vapor in an unsaturated polyester, *Polymer* 2000, 41, 2667–2676

[36] Cassie, A. B. D. Multimolecular absorption, *Transactions of the Faraday Society* 1945, 41, 450–458

[37] Hill, T. L. Statistical mechanics of multimolecular adsorption. I, *The Journal of Chemical Physics* 1946, 14, 263–267

[38] Hill, T. L. Statistical mechanics of multimolecular adsorption II. Localized and mobile adsorption and absorption, *The Journal of Chemical Physics* 1946, 14, 441–453

[39] Kohan, M. L. *Nylon Plastics Handbook*, Carl Hanser Verlag, 1995

[40] Feng, H.; Modeling of vapor sorption in glassy polymers using a new dual mode sorption model based on multilayer sorption theory, *Polymer* 2007, 48, 2988–3002

[41] Chang, M.-J.; Myerson, A. S.; Kwei, T. K. The effect of hydrogen bonding on vapor diffusion in water-soluble polymers, *Journal of Applied Polymer Science* 1997, 66, 279–291

[42] Thijs, H. M. L.; Becer, C. R.; Guerrero-Sanchez, C.; Fournier, D.; Hoogenboom, R.; Schubert, U. S. Water uptake of hydrophilic polymers determined by a thermal gravimetric analyzer with a controlled humidity chamber, *Journal of Materials Chemistry* 2007, 17, 4864–4871

[43] Ríos De Anda, A.; Fillot, L.-A.; Rossi, S.; Long, D.; Sotta, P. Influence of the sorption of polar and non-polar solvents on the glass transition temperature of polyamide 6,6 amorphous phase, *Polymer Engineering & Science* 2011, 51, 2129–2135

[44] Ríos De Anda, A.; Fillot, L. A.; Preda, F. M.; Rossi, S.; Long, D. R.; Sotta, P. Sorption and plasticization effects of ethanol–toluene–isooctane ternary mixtures in polyamide 6,6 and induced plasticization effects, *European Polymer Journal* 55 (2014) 199–209

[45] Sabard, M. Effet des conditions de procédé et des nanocharges sur la morphologie et les propriétés de sorption à l'eau et à l'ethanol (Ph.D. thesis), Université Claude Bernard Lyon I, 2011

[46] Sabard, M.; Gouanvé, F.; Espuche, E.; Fulchiron, R.; Seytre, G.; Fillot, L.-A.; Trouillet-Fonti, L. Influence of film processing conditions on the morphology of polyamide 6: Consequences on water and ethanol sorption properties, *Journal of Membrane Science* 2012, 415, 670–680

[47] Preda, F. M.; Dynamics of polyamide in the solid state in presence of solvents and in the molten state (Ph.D. thesis), Université Claude Bernard Lyon I, 2016

[48] Lasoski, S. W.; Cobbs, W. H. Moisture permeability of polymers. I. Role of crystallinity and orientation, *Journal of Polymer Science* 1959, 36, 21–33

[49] Michaels, A. S.; Parker, R. B. Sorption and flow of gases in polyethylene, *Journal of Polymer Science* 1959, 41, 53–71

[50] Michaels, A. S.; Bixler, H. J. Flow of gases through polyethylene, *Journal of Polymer Science* 1961, 50, 413–439

[51] Mitchell, J. B. O.; Price, S. L. On the relative strengths of amide...amide and amide...water hydrogen bonds, *Chemical Physics Letters* 1991, 180, 517–523

[52] Dixon, D. A.; Dobbs, K. D.; Valentini, J. J. Amide–water and amide–amide hydrogen bond strengths, *The Journal of Physical Chemistry* 1994, 98, 13435–13439

[53] Starkweather, H. W. The sorption of water by nylons, *Journal of Applied Polymer Science* 1959, 2, 129–133

[54] Starkweather, H. W. Some aspects of water clusters in polymers, *Macromolecules* 1975, 8, 476–479

[55] Lim, L.-T.; Britt, I. J.; Tung, M. A. Sorption and transport of water vapor in nylon 6,6 film, *Journal of Applied Polymer Science* 1999, 71, 197–206

[56] Picard, E.; Gérard, J.-F.; Espuche, E. Water transport properties of polyamide 6 based nanocomposites prepared by melt blending: On the importance of the clay dispersion state on the water transport properties at high water activity, *Journal of Membrane Science* 2008, 313, 284–295

[57] Morokuma, K. Molecular orbital studies of hydrogen bonds. III. C=O··· H–O hydrogen bond in $H_2CO\cdots H_2O$ and $H_2CO\cdots 2H_2O$, *The Journal of Chemical Physics* 1971, 55, 1236–1244

[58] Nagaraju, M.; Sastry, G. N. Effect of alkyl substitution on H-bond strength of substituted amide-alcohol complexes, *Journal of Molecular Modeling* 2011, 17, 1801–1816

2 Influence of Organic Solvent Sorption upon Water Infiltration and Molecular Mobility in Polyamides

Agustín Ríos de Anda, Florentina-Maria Preda,
Louise-Anne Fillot and Paul Sotta

2.1 Introduction

Polyamides are a family of semi-crystalline thermoplastic polymers that have emerged as good candidates to replace steel alloys in fuel tanks, thanks to their high mechanical performance and good barrier properties to toluene and isooctane, which are the main components of fuel [1]. However, new fuel formulations, colloquially known as biofuels, are increasingly used worldwide, and these contain a certain amount of ethanol. Thus, on the way to elaborating polyamide-based formulations with enhanced performance, it is essential to measure and rationalize barrier properties of polyamide films to various types of small molecules including both non-polar solvents (alkanes) and polar solvents (alcohols) [2]. Note also that atmospheric water may become an issue when dealing with ethanol.

Polyamides are semi-crystalline polymers possessing a crystalline, solvent-tight phase and an amorphous phase where solvent diffusion and sorption occur. The amorphous phase has a major impact on polyamide properties, since their crystalline ratio is relatively low (35–40 mass%) compared to other polymers such as polyethylene and polypropylene [3].

Polyamides possess a physical network of amide–amide hydrogen bonds (H-bonds) in both the amorphous and the crystalline phases [4–12]. The existence of this network, especially in the amorphous phase, presumably plays a major role in giving polyamides their good mechanical and thermal properties [13–21]. A general assumption about the plasticization effect of polar solvents is that they may disrupt a fraction of the interchain H-bonds originally present in the amorphous phase of the dry polymer, therefore increasing the molecular mobility or equivalently decreasing the T_g [22–24] This drop in T_g leads to a diminution of the mechanical properties, i.e. Young's modulus drops and polyamides become ductile [3, 25–28]. This assumption is particularly true for water. The effect of water on the properties of polyamides is well-documented [2, 4, 6, 12, 13, 21–46]. In addition, polyamides are particularly interesting to investigate because of the presence of amide groups, which act as specific H-bonding sorption sites [4, 12, 13, 38]; in the presence of water, H-bonds may indeed be formed between water and polar amide groups present along polymer chains [8, 47]. A recent study of the dynamics of water absorbed in polyamides has shown that the diffusion coefficient of water is considerably reduced with respect to bulk water, which indicates that strong interactions between water molecules and the polymer are formed [48]. Also, the motions of amide–water complexes have been discriminated [48].

In previous works [24, 50] we have studied the sorption of various polar, non-polar solvents, and ethanol–toluene–isooctane mixtures in bulk PA6,6. It was shown that an excess of both polar solvents (alcohols) and non-polar solvents (alkanes), at equilibrium, have a strong effect on the measured glass transition temperature (T_g) of the material. It was observed that polar solvents were absorbed in much larger amounts than non-polar solvents at sorption equilibrium. More unexpectedly, it was found that PA6,6 films immersed in pure toluene presented a significant decrease of the T_g, at small amounts of absorbed solvent. This effect was more important than that of water or ethanol, although the non-polar solvent toluene is not expected to have strong interactions (H-bonds) with the amide groups.

By reconsidering the sorption experiment, in this discussion we propose a simple explanation. We consider here the case in which the sorption cell is an open system, meaning that the solvent is in contact with airborne humidity. Applicative tests of barrier properties are often performed in such conditions. Water may thus infiltrate the solvent and the polymer until a thermodynamic equilibrium is reached. This chapter presents some general considerations on the sorption and plasticizing effect of ethanol–water mixtures, as well as that of pure polar and non-polar solvents on PA6,6 and PA6. Firstly, the thermodynamics of ethanol–water mixtures and their effect on the T_g values of such polyamides is discussed. Then the effect of "residual" water on pure polar and non-polar sorption and influence on the molecular mobility is detailed. Finally, it highlights the importance of performing sorption experiments in controlled conditions, in order to avoid water sorption as a side effect. It also emphasizes the importance of quantifying separately the amount of each sorbed species, for example by chromatographic techniques, as has been done previously [50].

Furthermore, in polyamides, it has been shown that the morphology (i. e. mainly the crystallinity ratio) may have a strong impact on sorption properties [36]. For extruded films, sorption properties were shown to depend significantly on the chill roll temperature, that is, on the thermal history of the samples, which in turns affects the crystallinity ratio and the crystalline phase structure [36]. This impact itself depends on the considered solvent, because the accessibility of the solvent to the whole amorphous phase may depend (for example) on the molecular size of the considered solvent. In the presence of non-ideal solvent mixtures, interactions between sorbed solvent molecules may induce non-trivial synergetic effects, leading to an overall sorption different from expected by simple addition of the sorbed amounts of each solvent as its corresponding activity in the mixture. However, to discriminate such non-trivial effects from the expected results, sorption results must be analyzed carefully. This is also needed to discriminate clearly the impact of the sample morphology on the sorption of various solvents or mixtures of solvents.

2.2 Experimental Methods

2.2.1 Materials

PA6, PA6,6, and PA6,10 pellets were provided by Solvay and contained no stabilizers or fillers. 500 µm-thick PA6 films and 100 µm-thick PA6,6 films were obtained by film-cast extrusion. PA6,10 plates 0.8 × 100 × 100 mm were obtained by injection-molding with a DEMAG H200-80T press. The plates were thinned at 300 µm using a planer. Prior to all experiments, the films were dried under vacuum for 24 h at 90 °C for PA6, 110 °C for PA6,6, and 80 °C for PA6,10. The films were introduced in thermo-sealed envelopes and kept in a desiccator to avoid any contact with air moisture. The glass transition temperature (T_g) and crystallinity ratio (χ_C) of each film, measured after drying, are:

- PA6 500 µm: T_g = 60 ± 2 °C, χ_C = 35%
- PA6,6 100 µm: T_g = 64 ± 2 °C, χ_C = 38%
- PA6,10 300 µm: T_g = 53 ± 2 °C, χ_C = 23%.

2.2.2 Sorption Experiments

Polymer films were immersed in liquid solvent in a non-hermetic recipient and removed regularly for weighing following a standard procedure: removal from sorption cell; pressing between absorbent paper; surface drying with compressed air; recording of the weight exactly 1 min after removal from sorption cell; re-immersion. The mass intake was recorded as a function of immersion time until equilibrium was reached. Measurements were carried out at temperatures between 25 and 66 °C.

Water–ethanol sorption was studied for PA6, PA6,6, and PA6,10. Moreover, for PA6, sorption experiments were performed with mixtures of ethanol and water of the following composition in volume percent: 100/0, 95/5, 85/15, 75/25, 50/50, 25/75, 15/85,

5/95 and 0/100. Furthermore, toluene sorption was studied for PA6,6 and PA6,10. Finally the sorption of a series of polar (alcohols) and non-polar (hydrocarbons) solvents of different sizes was undertaken on PA6,6. Their chemical structure is shown in Figure 2.1.

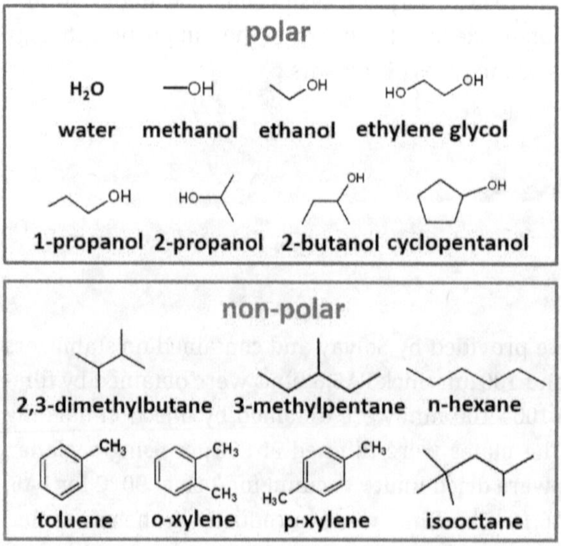

Figure 2.1 Molecular formulae of the studied solvents, including polar, non-polar aliphatic, and non-polar aromatic solvents

The studied polar solvents were water, methanol, ethanol, ethylene glycol, 1-propanol, 2-propanol, 2-butanol, and cyclopentanol. The alcohols can be separated into those that are fully miscible (hydrophilic), namely methanol, ethanol, ethylene glycol, 1-propanol, and 2-propanol, and those that are only partially miscible in water (amphiphilic), namely 2-butanol and cyclopentanol. The studied non-polar solvents can be classified as aliphatic hydrocarbons (2,3-dimethylbutane, 3-methylpentane, n-hexane, and isooctane) and aromatic hydrocarbons (toluene, *o*-xylene, and *p*-xylene). Deionized water and analytical-grade solvents were used. Moreover, polar solvents were carefully kept anhydrous using 3 Å molecular sieves as a desiccant.

The weight fraction of solvent Δm at time t is determined as a function of the mass of the polymer film at time t and the mass of the dry film m_0:

$$\Delta m(t) = \frac{m(t) - m_0}{m_0} \times 100\% \tag{2.1}$$

In addition, a useful quantity in these systems with specific sorption sites is the molar intake Δn, i. e. the ratio of solvent molecules to amide groups in the amorphous phase:

$$\Delta n = \frac{n_{\text{solvent}}}{n_{\text{amide groups}}} = \frac{\Delta m(t)}{100} \times \frac{M_{\text{PA}}}{M_{\text{solvent}}} \times \frac{1}{1-\chi_{\text{C}}} \tag{2.2}$$

Where:

M_{PA} = molecular weight/amide group (113 g/mol for PA6 and PA6,6 and 141 g/mol for PA6,10)

M_{solvent} = solvent molecular weight

χ_{C} = crystallinity ratio

2.2.3 Karl Fischer Titration

The water amount in polymer samples was determined by Karl Fischer titration with a Metrohm coulometer equipped with an oven. The solvent was desorbed by heating at high temperatures and carried in the measurement cell by a nitrogen flux. The output of the titration is the concentration of water in the sample in ppm.

2.2.4 Differential Scanning Calorimetry

A TA Instruments DSC Q2000 was used in standard mode (ramp 10 °C/min) to determine the melting point and χ_{C}, and in the modulated mode (MDSC) to determine the T_g of pristine and solvent-equilibrated polymers. MDSC was essential to this study because the glass transition of polyamide films could not be clearly determined in standard mode, as it is often the case in polyamides. A liquid nitrogen cooling system (LNCS) was used with a helium flow (25 mL/min) to have access to very low temperatures (down to –180 °C). Samples (7–15 mg) were placed in non-hermetic aluminium pans and heated from −150 °C to 150 °C at a heating rate of 3 °C/min with a temperature modulation of ±2 °C every 60 s. The solvent-swollen samples were transferred rapidly between the sorption cell and the DSC pans in order to limit solvent evaporation. The T_g was determined as the mid-height or inflexion point in the reversible heat flow.

2.3 Chemical Activities in Non-Ideal Solvent Mixtures

Consider a mixture of solvents A and B with molar compositions in the liquid χ_A and $\chi_B = 1 - \chi_A$. The saturated vapor pressure of each component as a function of temperature is described by the curves $p_A^{sat}(T)$ and $p_B^{sat}(T)$, as illustrated in Figure 2.2 for water and ethanol.

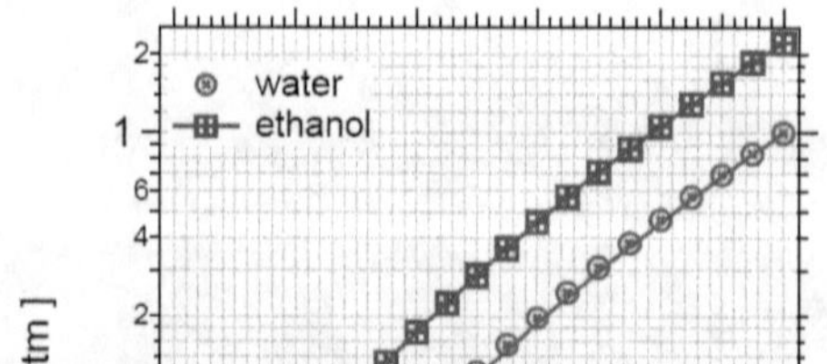

Figure 2.2 Saturated vapor pressures of water and ethanol as a function of temperature. Data points are from Lange's handbook [51]. For water, the fitting curve is described by $P_S(T) = P_S^0 \exp(-Q/T)$, where $P_S^0 = 1.14181 \times 10^6$ atm (1 atm = 1.012 bar), T is in K and Q = 5202.65 K [52]. For ethanol, the line is obtained based on the Antoine equation, $\log P^{sat}$ [mmHg] = 8.04494 – (1554.3/(T [°C] + 222.65)) [51]

For an ideal mixture, the activity of A is equal to the molar fraction in the liquid x_A (same for B). The composition of the vapor phase on top of the mixture at a temperature T (expressed by the molar fractions $y_A(T)$ and $y_B(T)$) can be calculated according to Equation 2.3 based on the partial vapor pressures of A ($x_A p_A^{sat}(T)$) and B ($x_B p_B^{sat}(T) = 1 - x_A p_A^{sat}(T)$):

$$y_A(T) = \frac{x_A p_A^{sat}(T)}{x_A p_A^{sat}(T) + (1 - x_A) p_B^{sat}(T)} = \frac{x_A p_A^{sat}(T)}{p_B^{sat}(T) + x_A \left(p_A^{sat}(T) - p_B^{sat}(T)\right)} \tag{2.3}$$

Thus, even for an ideal mixture, because of the difference between the saturated vapor pressures $p_A^{sat}(T)$ and $p_B^{sat}(T)$, the molar composition in the vapor $y_A(T)$ is not equal to the molar composition in the liquid $y_A(T)$. Furthermore, in this study, we are particularly interested in water–solvent mixtures, more specifically water–ethanol or water–toluene mixtures, which are *not ideal.*

2.4 Water–Ethanol Mixtures

For non-ideal mixtures, the activities are not equal to the molar fractions in the liquid phase and must be calculated in the vapor phase (assumed to be an ideal gas). Figure 2.3a shows the phase diagram of water–ethanol mixtures at 40 °C, in which the total vapor pressure P (the sum of water and ethanol vapor pressures) on top of the mixture is plotted as a function of the water molar content in the liquid. For a given total pressure, the molar fractions of water in the liquid (x_{water}) and vapor phase (y_{water}) are given by the blue and red curves respectively. The initial point of the curves at molar fraction zero is the saturated vapor pressure for pure ethanol at 40 °C ($P = 17.92$ kPa) and the final point at molar fraction one is the saturated vapor pressure for pure water at 40 °C ($P = 7.49$ kPa, see also Figure 2.2).

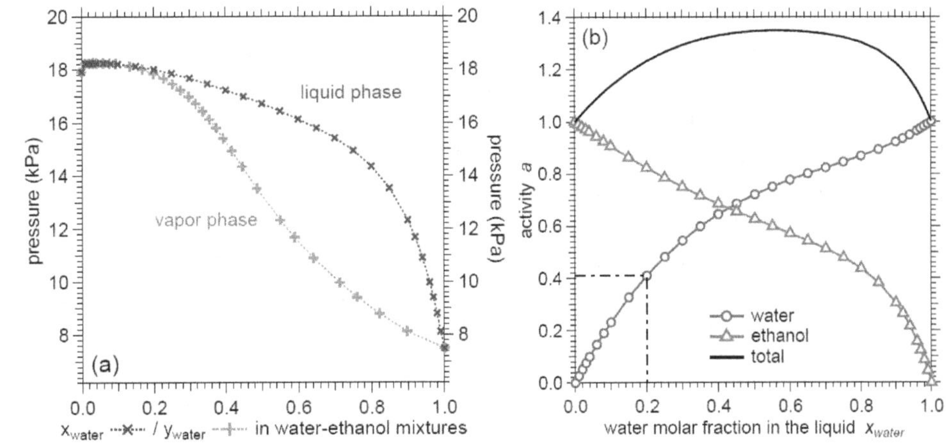

Figure 2.3 (a) Phase diagram of water–ethanol mixtures at 40 °C (x = molar fraction in the liquid, y = molar fraction in the vapor). Data from Cornell et al. [53]. (b) Activities of water (circles) and ethanol (triangles) as a function of the molar fraction of water in the liquid mixture, at 40 °C, together with their sum (black curve). At about RH40 (water activity $a_{water} = 0.4$) the liquid molar fraction is $x_{water} \approx 0.20$ (equivalent to a volume fraction of about 7 vol%), as indicated by dashed lines

The partial vapor pressure of water is then $P_{water} = y_{water} P_{tot}$, and that for ethanol is $P_{ethanol} = y_{ethanol} = y_{ethanol} P_{tot} = (1 - y_{water} P_{tot})$. The solvent activities are then $a_{water} = \left(P_{tot}/p^{sat}_{water}(T)\right)$ and $a_{ethanol} = \left(P_{tot}/p^{sat}_{ethanol}(T)\right)$. Both activities are plotted as a function of the molar content of water in the liquid mixture in Figure 2.3b, together with the total activity $a_{tot} = a_{water} + a_{ethanol}$. The sum of the activities is larger than unity, suggesting that there might be a synergistic effect of purely thermodynamic origin, i. e. the global uptake in the polyamide–solvent mixture may then increase compared to the pure solvents. Suppose that initially dry ethanol is kept under a humid atmosphere, for instance at RH50 ($a_{water} = 0.5$) at 40 °C. Then the liquid mixture will take water up to equilibrium at the same water activity in the liquid phase, as illustrated in Figure 2.3a. In this particular case for RH50, the molar fraction of water will be $x_{water} \approx 0.27$ in the liquid mixture, which corresponds to a weight fraction of 0.13 and a volume fraction of 0.1. The ethanol activity in the mixture is 0.78 (Figure 2.3b).

2.4.1 Sorption by Water–Ethanol Mixtures

In order to estimate the global intake (assuming that sorption of one solvent does not impact the other one), we need to know the sorption isotherms for each pure solvent. We shall discuss the cases of both PA6 and PA6,6. For PA6,6, sorption isotherms for water and ethanol at 40 °C were measured by dynamic vapor sorption (DVS) up to $a = 0.9$ in a previous study [54, 55], and are shown in Figure 2.4a. For PA6, sorption isotherms for water and ethanol at 40 °C were measured by Sabard et al. in extruded films [36] and are plotted in Figure 2.4b. These isotherms were also obtained by DVS up to $a = 0.9$. Points at $a = 1$ were obtained by immersing the samples in the liquid solvents, under ambient atmosphere (i. e. with some degree of humidity). It was shown that the chill roll temperature has a strong impact on sorption properties [36]. This was interpreted as due to the effect of the thermal history of the samples on the morphology, meaning both the crystalline ratio and the crystalline phase structure, which in turn have an impact on sorption [36].

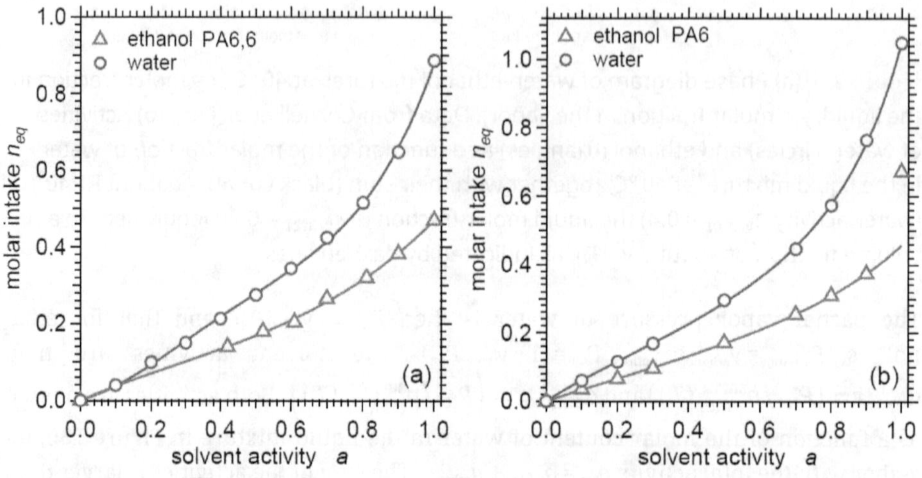

Figure 2.4 Sorption isotherms for water and ethanol at 40 °C in (a) PA6,6 [54, 55] and (b) PA6 [36]

Let us now consider the sorption when the samples are immersed in water–ethanol mixtures with various ethanol fractions. By combining sorption isotherms (Figure 2.4a and Figure 2.4b) and calculating the activities of each component in the mixtures (see Figure 2.3a), the total molar (or mass) intake can be computed, by assuming that the sorption of each component in the material is a function of its activity in the mixture only, independent of the presence of the other component in the material (that is, in the absence of any synergetic effect on sorption phenomena). Figure 2.5 shows the mass and molar intakes computed in this way.

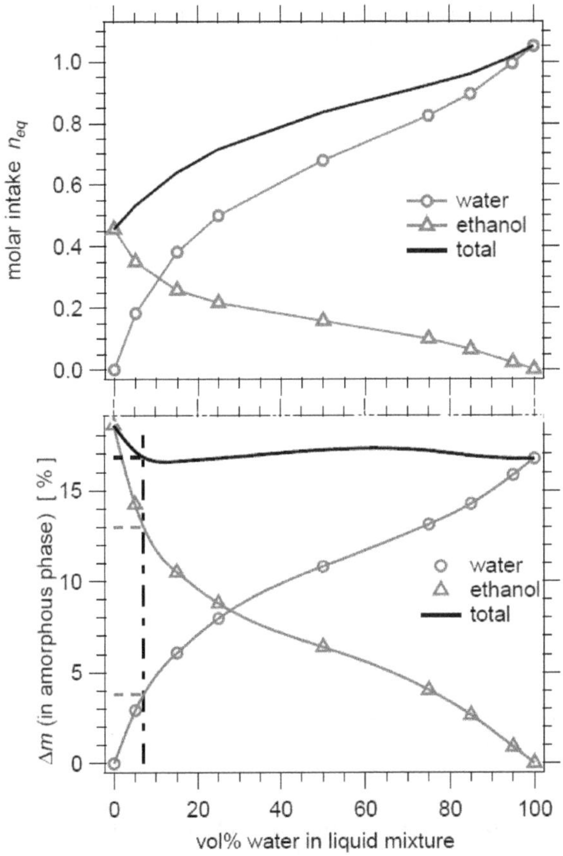

Figure 2.5 Water, ethanol and total molar intakes (top graph) and mass intakes (bottom graph) as a function of the water volume fraction in the mixtures, at 40 °C in PA6. These curves were obtained by combining sorption isotherms for PA6 (Figure 2.4 b and Figure 2.3b). Dashed lines indicate the intake values expected under an atmosphere with a humidity ratio of about 40%

Suppose now that the solvent mixture is placed under ambient atmosphere, with a humidity ratio (RH) of about 40%, meaning that the water activity in the gas phase on top of the mixture is fixed at about 0.4. Then, after reaching equilibrium, the mixture should absorb some amount of water, so as to equilibrate the water chemical potential in the atmosphere and in the mixture. As indicated in Figure 2.3b, this corresponds to a water molar fraction of about 0.2 (equivalent to a volume fraction of about 7 vol%).

From Figure 2.4a, the mass intake of water is 2.6 wt% at $a_{water} = 0.5$ and the mass intake of ethanol is 7.8 wt% at $a_{ethanol} = 0.78$, which would account for a total mass intake of 10.4 wt% in an ethanol–water mixture at equilibrium at 40 °C. The experimental mass intake is 11–12 wt%, slightly higher than the calculated one.

2.4.2 Effect on T_g

The effect on the T_g of the materials can then be estimated. For each solvent, T_g is related to the molar sorption as shown in Figure 2.6.

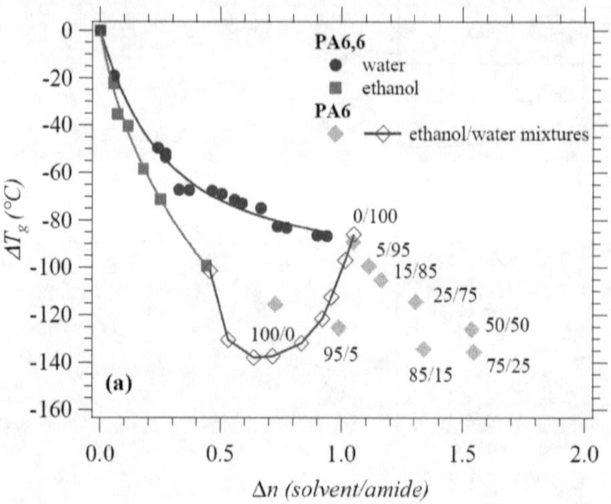

Figure 2.6 Shift in the T_g of PA6,6 in the presence of water, ethanol, and toluene and of PA6 in the presence of ethanol–water mixtures (volume fractions in the liquid are indicated on the figure next to the symbols), as a function of the molar intake Δn (solvent molecule/amide group). Since PA6,6 and PA6 have the same repeating unit, the effect of solvents on T_g should be close

A semi-quantitative value of the water intake in polyamide–solvent films can be obtained by Karl Fischer titration. The obtained quantities of water are reported in Table 2.1.

It can be seen in Figure 2.6 that in the case of ethanol in PA6,6, 1–2 wt% water was found in the films, out of the total 11–12 wt% ethanol–water absorbed. Although this measurement was done on films at 25 °C, this value is close to the estimated 2.6 wt% intake at 40 °C.

Table 2.1 Water and Ethanol Intakes in PA6 Films in Water–Ethanol Mixtures

Water-ethanol ratio [vol%]	Total Δm [wt%]	Δm_{water} [wt%]	$\Delta m_{ethanol}$ [wt%]	$\Delta n_{eq,water}$ (eq_{solv}/eq_{CONH})	$\Delta n_{eq,ethanol}$ (eq_{solv}/eq_{CONH})	ΔT_g MDSC [°C]	ΔT_g estimated [°C]
0/100	12.8	3.2	9.6	0.33	0.36	-116	-102
5/95	15.9	5.4	10.5	0.52	0.40	-125	-131
15/85	19.2	8.7	10.5	0.84	0.40	-134	-138
25/75	20.3	11.2	9.1	1.08	0.34	-136	-138
50/50	19.3	11.7	7.6	1.13	0.29	-126	-132
75/25	14.7	11.1	3.6	1.07	0.14	-115	-122
85/15	12.3	10.3	2.0	0.99	0.08	-105	-113
95/5	10.7	10.6	0.1	1.02	0.01	-99	-97
100/0	10.0	10.0	0.0	0.97	0.00	-89	-86

Water infiltration can be problematic, for instance if we want to understand the effect of solvents on the T_g of polyamide. Figure 2.7a illustrates the variation of the T_g of PA6 and PA6,6 at various molar intakes Δn of solvents, supposing that there is no water infiltration in ethanol or toluene. It can be observed that each class of molecule has its specific effect on T_g. Previously, an attempt to explain the T_g shift was made by reference to the different mechanisms of interaction (H-bonding for water and ethanol with amide groups, supposedly no interaction for toluene) [50]. However, the T_g shift in the presence of ethanol might have been overestimated, since some small percentage of water was probably present in the material. For example, a synergistic effect was observed in the shift of the T_g of polyamide in the presence of ethanol–water mixtures. The addition of only 5 vol% water to ethanol (95/5 mixture) results in an additional decrease in the T_g of 10 °C compared to pure ethanol (100/0). Therefore, it is crucial to take water infiltration into consideration.

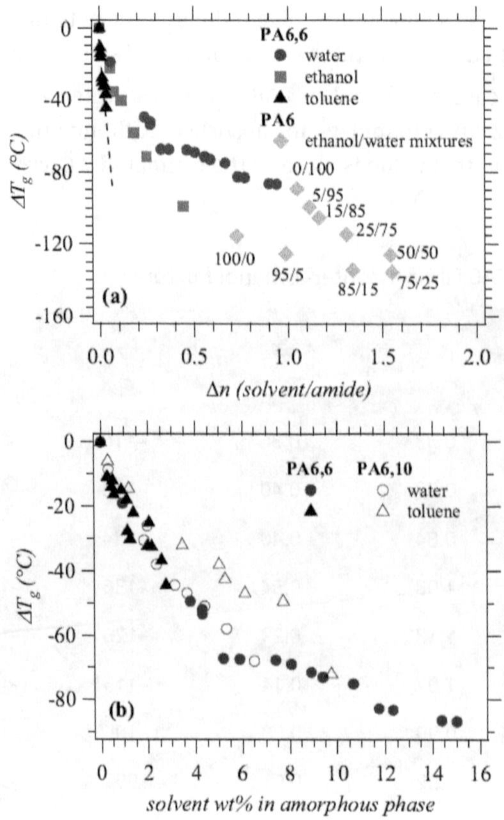

Figure 2.7 Shift in the T_g of PA6,6 and PA6,10 in the presence of water, ethanol, and toluene, as a function of the solvent mass intake in the amorphous phase

2.5 Dry-Assumed Solvents

As was discussed before, water infiltration can be problematic as it interacts strongly with polyamide and provokes a drop of polyamide T_g (i. e. an increase on its molecular mobility). The previous section describes how water was voluntarily added to ethanol to study the now plausible synergetic effect between these two solvents. In this section, the influence on solvent sorption and on the T_g of PA6,6 resulting from "residual" water present in pure solvents is discussed. It can be observed that each class of molecule has its specific effect on T_g.

However, according to the results shown in this work for water–ethanol mixtures, the solvent sorption as well as the T_g shift in the presence of pure alcohols and hydrocarbons might have been overestimated. Indeed, as was already stated for ethanol, "residual" water is also found in all alcohols (up to 1 vol%) and in hydrocarbons (up to 0.15 vol%) [51, 56]. As already stated before, introducing a desiccant into these solvents only prevents more water from being absorbed by these species [51, 56].

The amounts of pure polar and non-polar solvents absorbed by PA6,6 samples at equilibrium were measured, then the T_g (MDSC) of sorbed PA6,6 films were characterized. Table 2.2 summarizes the mass intakes Δm calculated molar intakes Δn_{eq} at equilibrium for such solvents according to Equation 2.1 and Equation 2.2. In this case, $M_{PA6,6}$ was taken as 113 g/mol and χ_C = 36%. Table 2.2 also includes the T_g values of the PA6,6 in contact with each solvent.

Table 2.2 Total Mass Δm and Molar Intakes Δn_{eq} at Sorption Equilibrium, Molar Volume V, Dipolar Moment μ and Relative Dielectric Constant ε_R of the Studied Solvents at Room Temperature [57, 58]

Solvent	Δm [wt%]	Δn_{eq} (eq_{solv}/eq_{CONH})	T_g [°C]	V [cm³/mol]	μ [× 10^{30} C·m]	ε_R
Dry	—	—	64	—	—	—
Water	9.6	0.97	-22	18.0	6.7	78.5
Methanol	13.2	0.74	-53	40.4	5.5	32.6
Ethanol	11.9	0.47	-35	58.3	5.7	24.3
Ethylene glycol	14.5	0.43	-36	55.8	6.7	37.0
1-Propanol	10.8	0.33	-30	74.7	5.6	20.1
2-Propanol	9.5	0.29	-10	76.3	5.5	18.3
2-Butanol	1.1	0.03	50	91.8	5.6	15.8
Cyclopentanol	1.0	0.02	51	90.6	5.5	6.9
Toluene	1.7	0.034	20	106.1	1.3	2.43
o-Xylene	2.9	0.049	13	120.5	1.5	2.57
p-Xylene	2.7	0.047	12	123.3	0.2	2.20
2,3-Dimethylbutane	2.0	0.042	25	132.8	—	—
3-Methylpentane	2.3	0.048	23	129.7	—	—
n-Hexane	1.2	0.026	40	130.7	0.3	1.89
Isooctane	1.9	0.030	35	165.1	0.0	2.10

Table 2.2 shows that total mass intakes Δm of water and relatively short (hydrophilic) alcohols in PA6,6 are comparatively high (ca. 10–15 wt%). The total mass intake is even larger for alcohols than for water. Conversely, amphiphilic alcohols (2-butanol

and cyclopentanol) and non-polar solvents have quite low total mass intakes when compared to polar solvents. Moreover, it appears that the non-polar solvents, despite their apparent low sorption, have a large plasticizer effect since they provoke a large T_g drop for PA6,6.

To extend these observations, sorption tests carried out on PA6,6 in contact with ethanol and toluene were compared to those undertaken on PA6,10 at 25 °C and 40 °C. Table 2.3 lists the mass intakes Δm for these solvents as well as the water intake measured by Karl Fischer experiments.

Table 2.3 Total Solvent Mass Intake at Equilibrium for Different Polyamides and the Estimated Amount of Water by Karl Fischer Titration (3 Å molecular sieves)

Polymer	Solvent	Sorption temperature [°C]	Total mass intake [wt%]	Water intake [wt%]	Solvent [wt%]
PA6,6	Toluene	25	0.3	0.5	0
	Toluene	25	0.4	0.5	0
	Toluene	25	1.7	2.1	0
	Toluene	40	0.5	0.3	0
	Ethanol	25	11.1	0.7	10.4
	Ethanol	25	11.3	1.0	10.3
	Ethanol	25	12.1	2.0	10.1
PA6,10	Toluene	25	1.6	0.3	1.3
	Toluene [a]	25	2.7	0.4	2.3
	Toluene [a]	25	6.2	1.7	4.6
	Toluene [a]	40	3.4	0.2	3.2
	Ethanol	25	7.1	0.4	6.7
	Ethanol [a]	25	10.7	0.3	10.4
	Ethanol [a]	25	11.7	0.3	11.4

[a] Solvent with dessicant (3 Å molecular sieves).

It is shown in Table 2.3 that there is "residual" water measured for the sorption of ethanol and toluene. Thus, to understand and comprehend the sorption and plasticization behavior of these solvents as well as any potential effect of "residual" water, these results will be thoroughly discussed in the following section.

2.5.1 Pure Polar Solvents

For this discussion we considered that all polar solvents, even with desiccant, had a "residual" water content of 1 vol%. In order to calculate the activity of water in such solvents, the same approach presented in the case of water–ethanol mixtures was used. Vapor–liquid equilibria data at room temperature in literature were solely found for water–methanol [59], water–ethylene glycol [60], and water–2-propanol [61]. These equilibria are shown in Figure 2.8.

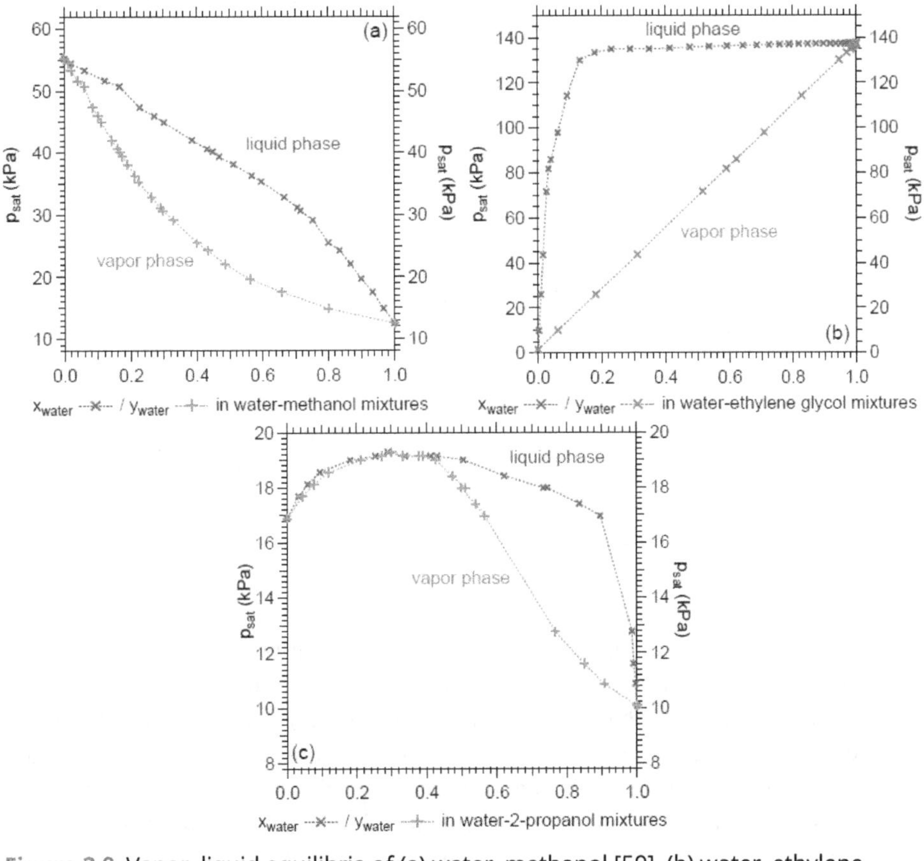

Figure 2.8 Vapor–liquid equilibria of (a) water–methanol [59], (b) water–ethylene glycol [60], and (c) water–2-propanol [61] mixtures at 25 °C (x = molar fraction in the liquid, y = molar fraction in the vapor)

From the vapor–liquid equilibria diagrams shown in Figure 2.8, the molar composition in the liquid phase x and the activity of water a_{water} were calculated. These values are listed in Table 2.4. The water sorption isotherm shown in Figure 2.4 was used to estimate the amount of this molecule absorbed by PA6,6 in contact with polar solvents according to the calculated activities a. The estimated values for water mass Δm_{water} and molar $\Delta n_{eq,\,water}$ intakes in such samples are also listed in Table 2.4 with the corrected mass $\Delta m_{solvent}$ and molar $\Delta n_{eq,\,solvent}$ intakes for polar solvents.

Table 2.4 Molar Fraction of Water in the Liquid Phase x_{water} in Polar Solvents, Water-Corrected Mass Δm (in wt%) and Molar Intakes Δn_{eq} (Referred to the Amorphous Phase in eq_{solv}/eq_{CONH}) at Sorption Equilibrium for Selected Polar Solvents

Solvent	x_{water}	a_{water}	Δm_{water} [wt%]	$\Delta m_{solvent}$ [wt%]	$\Delta n_{eq,water}$ (eq_{solv}/eq_{CONH})	$\Delta n_{eq,solvent}$ (eq_{solv}/eq_{CONH})
Methanol	0.018	0.044	0.22	12.98	0.02	0.72
Ethanol	0.025	0.063	0.31	11.59	0.03	0.44
Ethylene glycol	0.034	0.590	3.29	11.21	0.32	0.32
2-Propanol	0.029	0.079	0.40	9.10	0.02	0.27

It is seen in Table 2.4 that except for ethylene glycol, water intake remains relatively small and constant for the rest of the alcohols. One could expect that such a small amount of absorbed water would not significantly influence the sorption at equilibrium of polar solvents or their effect on PA6,6. However, it might be able to accelerate or "prime" diffusion of these solvents in these polymers by interacting with the amide groups through H-bonds. However, this remains a hypothesis that may be difficult to prove, as it is practically impossible to obtain completely waterless polar solvents under standard conditions. A solution to this would be to work under vacuum, where the ambient humidity would be absent.

The case of ethylene glycol could be explained by the extreme difference of vapor pressures between this molecule (0.22 psi) and water (19.93 psi) at room temperature [60]. In this system, water behaves as an extremely volatile liquid, meaning that a small amount of water in the liquid phase has a large activity in the system, as shown in Figure 2.8b. This would then prompt PA6,6 to absorb large amounts of water when in contact with ethylene glycol. The molar intake Δn_{eq} values listed in Table 2.4 show that in this system, for every molecule of ethylene glycol in PA6,6 a molecule of water is also absorbed. This differs from ethanol, which has a similar molecular size, but where, for every 100 molecules of this solvent present in PA6,6, only 7 molecules of water are absorbed by the polymer.

In this section and for further comparisons, the water-corrected molar intakes $\Delta n_{eq,solvent}$ for methanol, ethanol, ethylene glycol and 2-propanol were used. For the other polar solvents, the assumed molar intakes $\Delta n_{eq,total}$ were considered. In order to compare the different polar solvent intake values in PA6,6, Δn_{eq} was plotted as a function of the molecular size of each solvent. The parameter considered to define the molecular size of a given solvent is the molar volume $V = M/\rho$ (in cm^3/mol), where M is the molar mass and ρ the density of the solvent. The molecular intake Δn_{eq} plotted as a function of the molar volume V is shown in Figure 2.9 for such polar molecules.

Figure 2.9 shows that all data points for polar solvents seem to fall within a master curve. The molecular intake in PA6,6 decreases quite steeply as the molecular size increases. The sorption of such solvents might also be influenced not only by the solvent molecular volume, but also by the relative polarities of a given solvent molecule and PA6,6. Here, we propose to use the dielectric constant as a heuristic parameter to quantify the polarity of a solvent, which affects the interactions that it may have with polyamides.

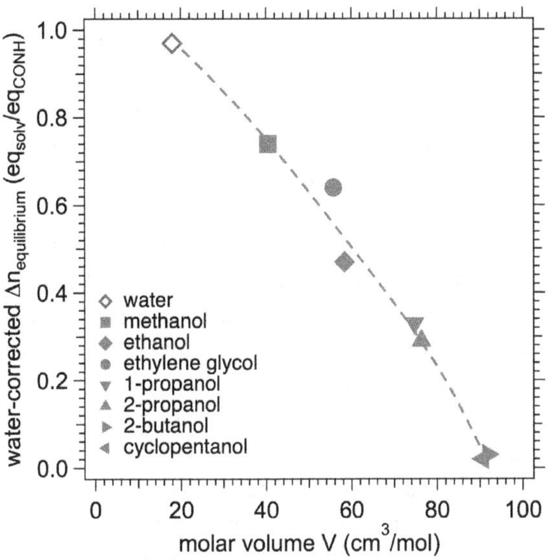

Figure 2.9 Molar intake at equilibrium Δn_{eq} as a function of the molar volume V for the polar solvents. The dashed line is a guide for the eye

The molar intakes of polar solvents were then plotted as a function of the relative dielectric constants ε_R (Figure 2.10a) or, alternatively, as a function of the molecular dipole moments μ (itself obtained from the dielectric constant ε_R of a molecule) divided by the molar volume V (Figure 2.10b). The ratio μ/V normalizes the value of the dipole moment of the hydroxyl function to the size of the solvent molecule. The literature values of ε_R and μ that have been used are tabulated in Table 2.2 [51, 56].

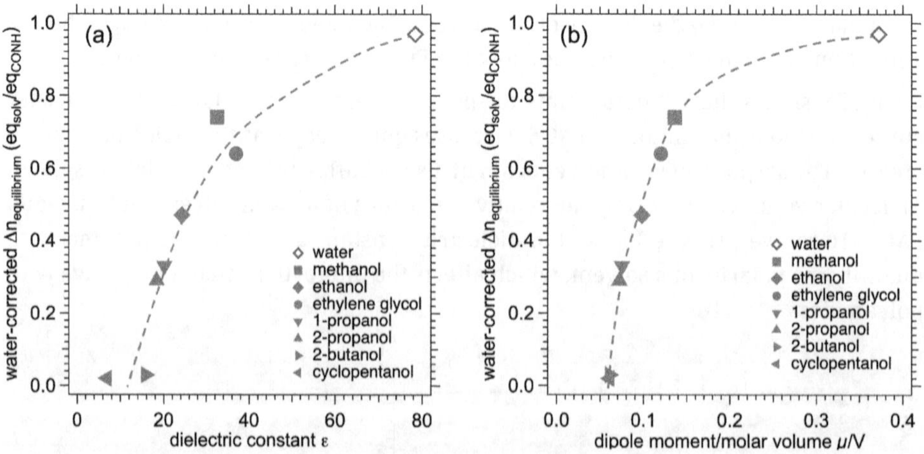

Figure 2.10 Molar intake at equilibrium Δn_{eq} for the studied polar solvents as a function of (a) their relative dielectric constant ε_R and (b) the ratio of their molar dipole moment μ over their molar volume V. The dashed lines are guides for the eye

It is observed in both Figure 2.10a and Figure 2.10b that there seems to exist a master curve or trend linking the polarity of a molecule to its intake in PA6,6. The molar intake is closely correlated to the polarity of the solvent. As the dielectric constant or the μ/V ratio increases, meaning that the molecule is more polar, the intake into PA6,6 increases as well. From these results, it can be concluded that the polarity of the solvent is the main factor conditioning its intake in PA6,6: the more polar a solvent is, the larger its intake will be into PA6,6. It has to be pointed out, however, that in this study the size and polarity of polar solvents cannot be decoupled from each other, as the increasing of size of these solvents corresponds to a decrease in the molecule polarity as well, since solvents having increasing alkyl functions in their structures were considered.

Dashed lines in Figure 2.10a and Figure 2.10b correspond to an adjustment of the data for mono-alcohols with a function of the form Δneq=A(εR-ε $\Delta n_{eq} = A(\varepsilon_R - \varepsilon_m)^{1/3}$ (in graph a) or $\Delta n_{eq} = B((\mu/v) - p_m)^{1/3}$, with $A = 0.243$, $\varepsilon_m = 15.8$, $B = 1.73$ and $p_m = 0.037$. This functional form is purely heuristic and the corresponding curves should merely be considered as guides for the eye. Ethylene glycol, which has two hydroxyl groups, falls within the general trend observed for mono-alcohols in Figure 2.9, Figure 2.10a, and Figure 2.10b, meaning that its bifunctionality does not influence its sorption in PA6,6.

The effect of polar solvents on the molecular mobility of PA6,6 was then assessed. In the case of water, it is already known that this molecule has a large effect on the molecular dynamics of this polymer. The most straightforward way to study the change in molecular mobility of the polymer is to measure its T_g. The measured T_g values for PA6,6 samples at swelling equilibrium by MDSC are reported in Table 2.2 and are plotted as a function of the molar intake at swelling equilibrium Δn_{eq} for all studied polar solvents in Figure 2.11. Curves showing the measured T_g as a function of the water or ethanol molecular intake in samples partially swollen with water or ethanol are also shown for comparison in Figure 2.11. The variation of T_g as a function of the molecular intake in samples partially swollen with water, ethanol, or toluene can be found in the literature [24].

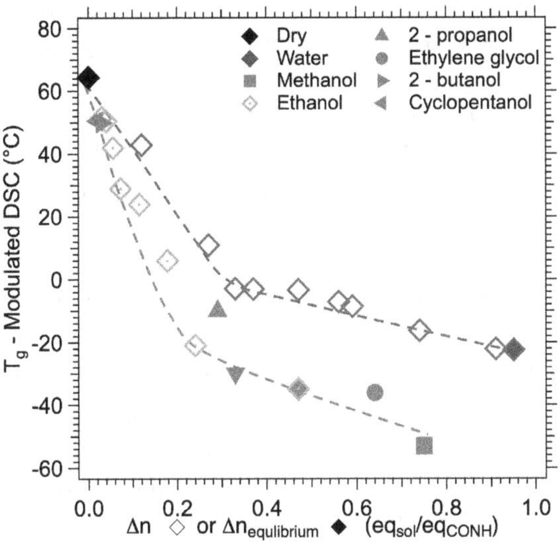

Figure 2.11 The T_g of PA6,6 as a function of molar intake at sorption equilibrium Δn_{eq} for samples saturated with polar solvents. Curves showing T_g as a function of the water or ethanol content in samples partially swollen with water or ethanol are taken from the literature [24] and are shown for comparison

It is observed in Figure 2.11 that, within experimental uncertainties, all data points for alcohols seem to fall on a master curve that exhibits the same trend as the curve corresponding to samples partially saturated with ethanol. Note that the point for ethylene glycol (which is a di-alcohol) falls on the same master curve as mono-alcohols. This suggests that similar plasticization mechanisms may be involved for all alcohol molecules. Water, however, does not follow the same trend. This solvent, which is the most absorbed one of all polar solvents, has a plasticizing effect at saturation comparable to medium size alcohols (i. e. 1- and 2-propanol). This means that, referred to a given molar intake, the relative plasticizing effect of water is weaker than that of alcohols. Moreover, it is important to notice that although a small amount of "residual" water is absorbed by PA6,6 samples in contact with alcohols, it does not seem to have a large influence on the T_g. To explain this, the following hypothesis is proposed.

Concerning water, it is generally accepted in the literature that this solvent is able to plasticize polyamides by bonding itself to the amide groups and replacing the existing interchain H-bond network of these polymers. Water is able to bond to the amide groups of PA6,6 because creating this bond corresponds to a gain of energy in the system, i. e. the energy of an amide–amide bond is ca. 25 kJ/mol whereas that of a water–amide bond is ca. 30 kJ/mol, as estimated by molecular modeling [38, 62–64]. Moreover, it has been proposed that one molecule of water that is already bonded to an amide group can form another bond with yet another molecule of water or with another amide group [9, 12, 30, 42–45, 65], somehow forming bridges between polymer chains. It might be possible then that for high concentrations of water in PA6,6, polymer chains are "linked" together through water bridges. In the case of alcohols being polar molecules, they might be able to directly interact with PA6,6 polymer chains by forming H-bonds between the hydroxyl functions of the alcohol molecules and the amide functions of PA6,6. The bonding between alcohols and the amide functions of PA6,6 is possible because this bond is stronger (ca. 30 kJ/mol for ethanol) than an amide–amide bond. This energy slightly drops when alkyl functions are added to the alcohol structure [66].

Since the studied alcohols have one hydroxyl function (except for ethylene glycol, which possesses two), one might think that by forming an H-bond with an amide function they would effectively disrupt the bonding between two amide functions, leading to a T_g drop of the polymer. Moreover, the alkyl functions of these solvents might also provoke a spacing of the polymer chains, stretching the H-bond network and leading to a reduction in the lifetime of such bonds and a subsequent T_g drop, as was mentioned for toluene. It would be possible then that the alcohol molecules have a "double" effect by both directly interacting with the H-bond network and also provoking a spacing of PA6,6 chains through steric effects. Both effects would effectively counteract any influence of absorbed "residual" water in PA6,6.

2.5.2 Pure Non-Polar Solvents

As mentioned before, it was observed that toluene penetrated in small quantities in PA6,6 but apparently had a very important effect on the glass transition temperature of the polyamide films [24,50]. This behavior could not be accurately accounted for at the time, so more investigations were done during the current work. The case of non-polar solvents containing "residual" water is particularly interesting, since non-polar solvents are assumed to not be able to mix with water. However, percentages of water ranging from 0.05 vol% for isooctane, 0.1 vol% for toluene, and 0.15 vol% for n-hexane, *o*-xylene, and *p*-xylene are found at equilibrium in these solvents under a relative humidity of 100% [51, 56]. In this work, these values were taken to estimate the molar fraction of water in the liquid phase x_{water} in non-polar solvents. These values are listed in Table 2.5.

Table 2.5 Molar Fraction of Water in the Liquid Phase x_{water} inNon-Polar Solvents, Water- Corrected Mass Δm_{water}, and Molar Intakes Δn_{eq} at Sorption Equilibrium for Selected Non-Polar Solvents

Solvent	a_{water}	x_{water}	Δm_{water} [wt%]	$\Delta m_{solvent}$ [wt%]	$\Delta n_{eq,water}$ (eq_{solv}/eq_{CONH})	$\Delta n_{eq,solvent}$ (eq_{solv}/eq_{CONH})
Toluene	0.5	0.050	1.7	0.0	0.17	0.0
o-Xylene	0.5	0.071	2.9	0.0	0.28	0.0
p-Xylene	0.5	0.071	2.7	0.0	0.26	0.0
n-Hexane	0.5	0.068	1.2	0.0	0.12	0.0
Isooctane	0.5	0.031	1.9	0.0	0.19	0.0

At RH50, i. e. for an activity of water $a_{water} = 0.5$, the molar fraction of water in non-polar solvents x_{water} can be estimated as shown in Figure 2.12.

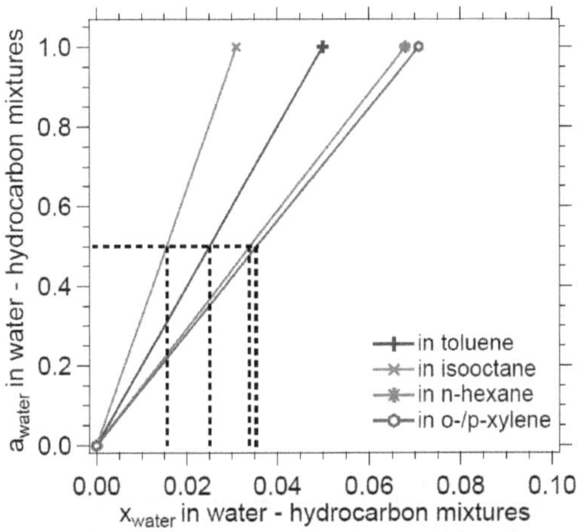

Figure 2.12
Activities of water and selected hydrocarbons as a function of the molar fraction of water in the liquid mixture at 25 °C

Although the amount of water x_{water} is extremely small, its activity is equal to the external activity ($a_{water} = 0.5$) at equilibrium, which will drive its sorption in the polymer. Since the sorption isotherms of non-polar solvents are not available, the total amount of solvent cannot be estimated. Based on the sorption isotherms of water [54], however, 2.6 wt% of this molecule was absorbed at $a_{water} = 0.5$ by PA6,6. Moreover, Karl Fischer titrations were carried out on PA6,6 films that were put in contact with toluene that was, a priori, absorbed between 0.3 and 1.7 wt% (see Table 2.3). However these titrations showed that these absorbed amounts corresponded to water in the films. This would mean that toluene was not in fact absorbed by PA6,6. Taking this results, as well as the sorption of pure water at $a_{water} = 0.5$, it could be deduced that for the rest of the non-polar solvents, their intake in PA6,6 actually corresponds to absorbed "residual" water, as in the case of toluene. From this assumption, the corrected molar intake $\Delta n_{eq,water}$ in PA6,6 samples in contact with hydrocarbons was calculated. These values are listed in Table 2.5.

The molecular intake of "residual" water $\Delta n_{eq,water}$ for PA6,6 films in contact with non-polar solvents has been plotted as a function of the non-polar solvents relative dielectric constant ε_R in Figure 2.13. Data for 2-butanol and cyclopentanol is shown for comparison. Figure 2.13 shows that the molar intakes for non-polar solvents seem to decrease linearly with their dielectric constant ε_R. "Residual" water in aromatic solvents seem to have a consistently higher intake than that in aliphatic ones. This can be correlated to a slightly higher polarizability of aromatic solvents. This result indicates that the amount of "residual" water found at equilibrium in non-polar solvents depend on the polarizabilities of these solvents. In other words, the more polarizable a non-polar molecule is, the more "residual" water it will contain.

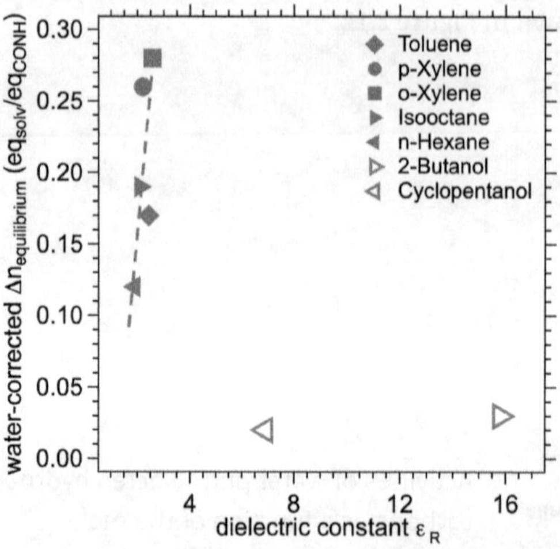

Figure 2.13 Molar intake of "residual" water at equilibrium $\Delta n_{eq,water}$ for PA6,6 films in contact with non-polar solvents, and for 2-butanol and cyclopentanol as a function of their relative dielectric constant ε_R. Same as in Figure 2.11, emphasized to show non-polar solvents. The dashed line is a guide for the eye

It can be ascertained now that the plasticizing effect of non-polar solvents, is actually due to the presence of "residual" water being absorbed by PA6,6. Figure 2.14 shows two different set of values of the variation of the T_g of PA6,6: the first as a function of the solvent's molar intake $\Delta n_{eq,solvent}$ as if the T_g drop would be due to the actual sorption of non-polar solvents, the second being the effect of absorbed "residual" water $\Delta n_{eq,water}$. For the sake of comparison, the plasticization curve of pure water is also plotted in Figure 2.14.

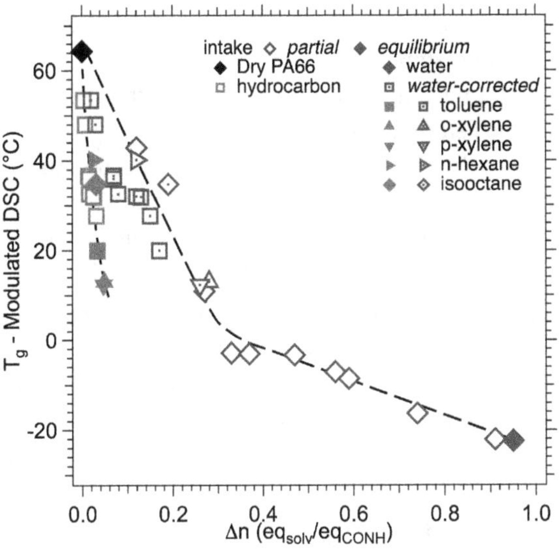

Figure 2.14 The T_g of PA6,6 as a function of molar intake at sorption equilibrium $\Delta n_{eq,solvent}$ as if the absorbed solvents were the non-polar series (red markers) and as if the absorbed solvents were "residual" water in non-polar solvents $\Delta n_{eq,water}$ (purple markers). The plasticization curve of water is shown for comparison (blue markers). Dashed lines are guides for the eyes

Figure 2.14 shows that the T_g values for absorbed "residual" water in PA6,6 films in contact with non-polar solvents follows the same trend as that of pure water. These results confirm that "residual" water and not non-polar solvents are absorbed by PA6,6. Thus the plasticization effect observed for PA6,6 films in contact with such molecules is solely due to the presence of water in these solvents that is absorbed by this polymer.

The sorption curves of toluene in PA6,6 or PA6,10 films were recorded at different temperatures by immersion in a non-hermetic sorption cell containing initially anhydrous toluene. The obtained curves are shown in Figure 2.15.

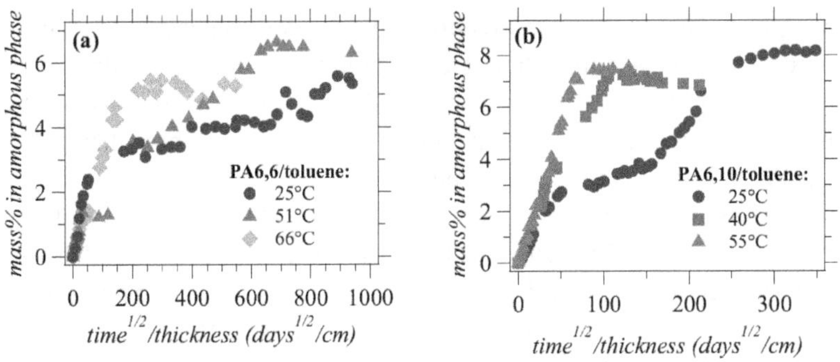

Figure 2.15 Sorption kinetics curves of toluene in (a) PA6,6 and (b) PA6,10 at different temperatures

At the first equilibrium plateau, small quantities (1–3 wt%) of solvent are absorbed in the amorphous phase. However, a second step of sorption equilibrium appears in all cases at long immersion times (several months). The solvent mass intake reaches values up to 8 wt%. Moreover, it was observed that the sorption kinetics curves varied significantly from one polymer film to another in a series of repeated experiments or if the measurements were done in winter or summer (variability in air humidity). In addition, the first part of the curves up to the first equilibrium corresponds roughly with the sorption curves of water at $a_{water} = 0.5$ measured for these systems.

Karl Fischer measurements provided estimations of the water content for these systems, as can be seen in Table 2.3. In PA6,6 with toluene, the Karl Fischer titration of water corresponds roughly to the total amount of solvent in the polymer up to 2 wt%. Therefore, it would seem that only water enters the polymer. This hypothesis is confirmed by the curve in Figure 2.14, in which the T_g shift corresponding to a supposed mass intake of toluene superimposes perfectly on the T_g shift for the mass intake of water at 25 °C. Therefore, the important effect seen on T_g in Figure 2.14 would be only an artifact from the calculation with Equation 2.2, which takes into account the molar mass of toluene (92 g/mol vs. 18 g/mol for water).

In PA6,10 with toluene, increasing amounts of water are found in the polymer as the total solvent intake increases (Table 2.3, toluene 25 °C in PA6,10). It would seem that water and toluene penetrate simultaneously in PA6,10. The T_g shift in Figure 2.7b confirms this hypothesis: the curve of toluene does not superimpose onto that of water. As in the case of ethanol, there might be a synergistic effect of the mixture. It should be highlighted that the sorption mechanism is different in PA6,6 and PA6,10. Indeed, in PA6,6, water penetrated first, followed by toluene, whereas in PA6,10 both solvents penetrate simultaneously. The different polarity of the polymers could be at the origin of this behavior.

2.6 Conclusion

In this chapter we have presented thermodynamic calculations to prove water infiltration in polyamide–solvent systems. Ethanol and water can be mixed in all proportions and their phase diagram allows us to calculate the amount of water at equilibrium with ethanol at 40 °C in a sorption cell in contact with ambient atmosphere. The sorption isotherms of water provide then the mass percentage intake of water in the polyamide film. This quantity is in agreement with Karl Fischer measurements. A synergistic effect is evidenced in these systems from simple thermodynamics, i. e. the amount of absorbed solvents from the mixture is higher than the absorbed amount of individual solvents. The same synergistic effect was evidenced on the T_g shift in the presence of water–ethanol mixtures.

Toluene and water are not miscible in all proportions because of their difference in polarity. Under atmospheric conditions, toluene will absorb 0.05 wt% water [51]. The sorption kinetics curves for toluene (+ water) show a peculiar behavior, and the effect on T_g is very important for small amounts of (hypothetically) absorbed toluene. This effect could be mainly attributed to the water infiltration. A different sorption behavior was highlighted in PA6,6 and PA6,10.

In addition, there might exist "dynamic" synergistic effects related to the influence of one solvent on the dynamics of the polymer (or the H-bond network), which would in turn influence the sorption of the other solvent (or conversely). However, quantification of this effect requires first to take into account the non-ideal nature of solvent mixtures.

In conclusion, it is very important, firstly to study solvent sorption (other than water) in closed systems like that provided by a dynamic vapor sorption device, and secondly to take into account the phase diagram in non-ideal mixtures. These conclusions are valid for solvent sorption in all polymer matrices.

References

[1] Yeh, J.-T.; Chang, S.-S.; Yao, H.-T.; Chen, K.-N.; Jou, W.-S. The permeation resistance of polyethylene, polyethylene/polyamide and polyethylene/modified polyamide blown tubes against unleaded gasoline. *Journal of Materials Science* 2000, 35, 1321–1330

[2] Skirrow, G.; Young, K. Sorption, diffusion and conduction in polyamide-penetrant systems: 1. Sorption phenomena. *Polymer* 1974, 15, 771–776

[3] Kohan, M. *Nylon Plastics Handbook*. Hanser, 1995

[4] Starkweather, H. The sorption of water by Nylons. *Journal of Applied Polymer Science* 1959, 2, 129

[5] Starkweather, H., Whitney, J.; Johnson, D. Crystalline order in Nylon 66. *Journal of Polymer Science Part A* 1963, 1, 715–723

[6] Starkweather, H. Some aspects of water clusters in polymers. *Macromolecules* 1975, 8, 476

[7] Starkweather, H.W.; Barkley, J.R. The effect of water on the secondary dielectric relaxations in Nylon 66. *Journal of Polymer Science: Polymer Physics Edition* 1981, 19, 1211–1220

[8] Skrovanek, D.J.; Howe, S.E.; Painter, P.C.; Coleman, M.M. Hydrogen bonding in polymers: Infrared temperature studies of an amorphous polyamide. *Macromolecules* 1985, 18, 1676–1683

[9] Murthy, N.; Stamm, M.; Sibilia, J.; Krimm, S. Structural changes accompanying hydration in Nylon 6. *Macromolecules* 1989, 22, 1261–1267

[10] Hutchison, J.L.; Murthy, N.S.; Samulski, E.T. Deuterium NMR studies of water in oriented Nylon 6 fibers. *Macromolecules* 1996, 29, 5551–5557

[11] Murthy, N.; Akkapeddi, M. Analysis of lamellar structure in semicrystalline polymers by studying the absorption of water and ethylene glycol in nylons using small-angle neutron scattering. *Macromolecules* 1998, 31, 142–152

[12] Murthy, N.S. Hydrogen bonding, mobility, and structural transitions in aliphatic polyamides. *Journal of Polymer Science Part B* 2006, 44, 1763–1782

[13] Garcia, D.; Starkweather, H. Hydrogen bonding in Nylon 66 and model compounds. *Journal of Polymer Science: Polymer Physics Edition* 1985, 23, 537–555

[14] Starkweather, H.Xenopoulos, A.; Wunderlich, B. Thermodynamic properties of liquid and semicrystalline linear aliphatic polyamides. *Journal of Polymer Science Part B* 1990, 28, 2271–2290

[15] Huang, J. J.; Keskkula, H.; Paul, D. R. Rubber toughening of an amorphous polyamide by functionalized SEBS copolymers: Morphology and Izod impact behavior. *Polymer* 2004, 45, 4203–4215

[16] Huang, J. J.; Keskkula, H.; Paul, D. R. Comparison of the toughening behavior of Nylon 6 versus an amorphous polyamide using various maleated elastomers. *Polymer* 2006, 47, 639–651

[17] Huang, J. J.; Paul, D. R. Comparison of fracture behavior of nylon 6 versus an amorphous polyamide toughened with maleated poly (ethylene-1-octene) elastomers. *Polymer* 2006, 47, 3505–3519

[18] Schroeder, L. R.; Cooper, S. L. Hydrogen bonding in polyamides. *Journal of Applied Physics* 1976, 47, 4310–4317

[19] Triggs, N. E.; Bonn, R. T.; Valentini, J. J. Do all solid amides hydrogen bond? Raman evidence to the contrary. *Journal of Physical Chemistry* 1993, 97, 5535–5540

[20] Miura, H.; Hirschinger, J.; English, A. D. Segmental dynamics in the amorphous phase of nylon 66: solid state deuterium NMR. *Macromolecules* 1990, 23, 2169–2182

[21] Prevorsek, D. C.; Butler, R. H.; Reimschuessel, H. K. Mechanical relaxations in polyamides. *Journal of Polymer Science Part A-2: Polymer Physics* 1971, 9, 867–886

[22] Khanna, Y. P.; Kuhn, W. P.; Sichina, W. J. Reliable measurements of the Nylon 6 glass transition made possible by the new dynamic DSC. *Macromolecules* 1995, 28, 2644–2646

[23] Lim, L.; Britt, I.; Tung, M. Sorption and transport of water vapor in Nylon 6,6 film. *Journal of Applied Polymer Science* 1999, 71, 197

[24] Ríos de Anda, A.; Fillot, L. A.; Rossi, S.; Long, D.; Sotta, P. Influence of the sorption of polar and non-polar solvents on the glass transition temperature of polyamide 6,6 amorphous phase. *Polymer Engineering & Science* 2011, 51, 2129–2135

[25] Ríos de Anda, A. Influence of the solvent sorption, additivation, and chemical modification on the molecular mobility dynamics of polyamide 6,6 amorphous phase and its consequences on the tensile and impact strength properties of this polymer [Thesis], Université de Lyon, 2012

[26] Ríos de Anda, A.; Fillot, L.-A.; Long, D. R.; Sotta, P. Influence of the amorphous phase molecular mobility on impact and tensile properties of polyamide 6,6. *Journal of Applied Polymer Science* 2016, 133

[27] Gaymans, R. J.; Borggreve, R. J. M.; Spoelstra, A. B. Ductile transition in nylon–rubber blends: Influence of water. *Journal of Applied Polymer Science* 1989, 37, 479–486

[28] Miri, V.; Persyn, O.; Lefebvre, J.-M.; Seguela, R. Effect of water absorption on the plastic deformation behavior of Nylon 6. *European Polymer Journal* 2009, 45, 757–762

[29] Auerbach, I.; Carnicom, M. Sorption of water by Nylon 6,6 and Kevlar 29. *Journal of Applied Polymer Science* 1991, 42, 2417

[30] Puffr, R.; Sebenda, J. On the structure and properties of polyamides. XXVII. The mechanisms of water sorption in polyamides. *Journal of Polymer Science Part C* 1967, 16, 79–93

[31] Reimschuessel, H. K. Relationship on the effect of water on glass transition temperature and Young's modulus of Nylon 6. *Journal of Polymer Science Part B* 1978, 16, 1229–1236

[32] Laredo, E.; Hernández, M. Moisture effect on the low- and high-temperature dielectric relaxations in Nylon-6. *Journal of Polymer Science Part B* 1997, 35, 2879

[33] Laredo, E.; Grimau, F.; Sánchez, F.; Bello, A. Water absorption effect on the dynamic properties of Nylon-6 by dielectric spectroscopy. *Macromolecules* 2003, 36, 9840

[34] Pathmanathan, K.; Cavaillé, J.-Y.; Johari, G. The dielectric properties of dry and water-saturated Nylon-12. *Journal of Polymer Science Part B* 1992, 30, 341

[35] Le Huy, H.; Rault, J. Remarks on the α and β transitions in swollen polyamides. *Polymer* 1994, 35, 136

[36] Sabard, M.; Gouanvé, F.; Espuche, E.; Fulchiron, R.; Seytre, G.; Fillot, L.-A.; Trouillet-Fonti, L. Influence of film processing conditions on the morphology of polyamide 6: Consequences on water and ethanol sorption properties. *Journal of Membrane Science* 2012, 415, 670–680

[37] Y. P., K.; Han, P.; Day, E. New developments in the melt rheology of Nylons. I: Effect of moisture and molecular weight. *Polymer Engineering & Science* 1996, 36, 1745–1754

[38] Jorgensen, W.; Swenson, C. Optimized intermolecular potential functions for amides and peptides. Hydration of amides. *Journal of the American Chemical Society* 1985, 107, 1489

[39] Camacho, W.; Hedenqvist, M. S.; Karlsson, S. Near infrared (NIR) spectroscopy compared with thermogravimetric analysis as a tool for on-line prediction of water diffusion in polyamide 6,6. *Polymer International* 2002, 51, 1366–1370

[40] Hernandez, R. J.; Giacin, J. R.; Grulke, E. A. The sorption of water vapor by an amorphous polyamide. *Journal of Membrane Science* 1992, 65, 187–199

[41] Razumovskii, L. P.; Markin, V. S.; Zaikov, G. Y. Sorption of water by aliphatic polyamides. Review. *Polymer Science USSR* 1985, 27, 751–768

[42] Iwamoto, R.; Murase, H. Infrared spectroscopic study of the interactions of Nylon-6 with water. *Journal of Polymer Science Part B: Polymer Physics* 2003, 41, 1722–1729

[43] Rastogi, S.; Terry, A. E.; Vinken, E. Dissolution of hydrogen-bonded polymers in water: A study of Nylon-4,6. *Macromolecules* 2004, 37, 8825–8828

[44] Vinken, E.; Terry, A. E.; van Asselen, O.; Spoelstra, A. B.; Graf, R.; Rastogi, S. Role of superheated water in the dissolution and perturbation of hydrogen bonding in the crystalline lattice of polyamide 4,6. *Langmuir* 2008, 24, 6313–6326

[45] Dixon, D. A.; Dobbs, K. D.; Valentini, J. J. Amide–water and amide–amide hydrogen bond strengths. *Journal of Physical Chemistry* 1994, 98, 13435–13439

[46] Feyereisen, M. W.; Feller, D.; Dixon, D. A. Hydrogen bond energy of the water dimer. *Journal of Physical Chemistry* 1996, 100, 2993–2997

[47] Goudeau, S.; Charlot, M.; Vergelati, C.; Müller-Plathe, F. Atomistic simulation of the water influence on the local structure of PA66. *Macromolecules* 2004, 37, 8072–8081

[48] Laurati, M.; Sotta, P.; Long, D. R.; Fillot, L. A.; Arbe, A.; Alegría, A.; Embs, J. P.; Unruh, T.; Schneider, G. J.; Colmenero, J. Dynamics of water absorbed in polyamides. *Macromolecules* 2012, 45, 1676–1687

[49] Mourglia-Seignobos, E.; Long, D. R.; Odoni, L.; Vanel, L.; Sotta, P.; Rochas, C. Physical mechanisms of fatigue in neat polyamide 6,6. *Macromolecules* 2014, 47, 3880–3894

[50] Ríos De Anda, A.; Fillot, L. A.; Preda, F. M.; Rossi, S.; Long, D. R.; Sotta, P. Sorption and plasticization effects of ethanol–toluene–isooctane ternary mixtures in polyamide 6,6 and induced plasticization effects. *European Polymer Journal* 2014, 55, 199–209

[51] Dean, J.; Lange, N. *Lange's Handbook of Chemistry*. McGraw-Hill, 1992

[52] Landau, L.; Lifshitz, E. *Physique Statistique* [4th edition]. Ellipses, 1995

[53] Cornell, L.; Montonna, R. Studies in distillation. II. Liquid-vapor equilibria in the systems ethanol–water, methanol–water and acetic acid–water. *Industrial and Engineering Chemistry* 1933, 25, 1331–1335

[54] Preda, F.-M.; Alegría, A.; Bocahut, A.; Fillot, L.-A.; Long, D. R.; Sotta, P. Investigation of water diffusion mechanisms in relation to polymer relaxations in polyamides. *Macromolecules* 2015, 48, 5730–5741

[55] Preda, F. Dynamics of polyamide in the solid state in presence of solvents and in the molten state [Thesis]. Université de Lyon, 2016

[56] Perry, R. H. *Chemical Engineers' Handbook* [7th edition]. McGraw-Hill, 1997. (b) Hodgman, C. *CRC Handbook of Chemistry and Physics* [44th edition]. CRC Press, 1962

[57] Lechner, M. D.; Wohlfarth, C. *Static Dielectric Constants of Pure Liquids and Binary Liquid Mixtures*. Springer, 2008

[58] Kurihara, K.; Minoura, T.; Takeda, K.; Kojima, K. Isothermal vapor–liquid equilibria for methanol+ethanol+water, methanol+water, and ethanol+water. *Journal of Chemical and Engineering Data* 1995, 40, 679–684

[59] Kireev, V.; Popov, A. Equilibria in mixtures of liquids and solutions. I. Boiling points and the composition of aqueous solutions of ethylene glycol and ethylene oxide. *Zhurnal Pikladnoi Khimii* 1934, 7, 489–494

[60] Barr-David, F.; Dodge, B. F. Vapor–liquid equilibrium at high pressures. The systems ethanol–water and 2-propanol–water. *Journal of Chemical and Engineering Data* 1959, 4, 107–121

[61] Klotz, I. M.; Franzen, J. S. Hydrogen bonds between model peptide groups in solution. *Journal of the American Chemical Society* 1962, 84, 3461–3466

[62] Gill, S.; Noll, L. Calorimetric study of association of diketopiperazine in water. *Journal of Physical Chemistry* 1972, 76, 3065–3068

[63] Chemical relaxation as a mechanistic probe of hydrogen bonding. Thermodynamics and kinetics of lactam isoassociation in nonpolar solvents. *The Journal of Physical Chemistry* 1974, 78, 2341–2348

[64] Eberhardt, E. S.; Raines, R. T. Amide–amide and amide–water hydrogen bonds: Implications for protein folding and stability. *Journal of the American Chemical Society* 1994, 116, 2149–2150

[65] Nagaraju, M.; Narahari Sastry, G. Effect of alkyl substitution on H-bond strength of substituted amide–alcohol complexes. *Journal of Molecular Modeling* 2011, 17, 1801–1816

3 Diffusion of Water and Oils in Thermosets and Thermoset Composites

Simon Bard, Martin Demleitner, Fabian Hübner,
Volker Altstädt and Holger Ruckdäschel

3.1 Introduction

Thermosets are widely used in fiber-reinforced composites and as coatings and protective layers in various applications. A main difference compared to thermoplastic materials is that the mobility of the thermoset polymer chains is reduced due to the crosslinked nature of the material. This limits the sorption of gases and liquids, and enhances the mechanical behavior at higher temperatures. Amongst other applications, in electric and fossil-fuel engines, they are also in contact with different oils used for cooling. The requirements of thermosets in these environments to withstand higher temperatures is now rising, especially with the emergence of electric engines. With an increase in the temperature, the sorption and diffusion of cooling fluids may also increase. The book chapter summarizes research on thermoset composite and coating applications. A focus is placed on the diffusion of fluids, namely water and "Skydrol", an aviation hydraulic fluid. Indeed, gas diffusion among these materials has been extensively discussed.

Sorption and diffusion of liquids and gases in thermosets is of considerable interest both for academic research and for industrial users. Very often, the effect is visible after several years of the part being in use. Sorption and diffusion are closely related to temperature, surrounding medium, network structure, mechanical loads, fillers, and (in the case of composites) the fiber type and orientation. The complex interplay between different parameters needs to be understood well in order to predict the long-term behavior.

The diffusion of water is especially of interest in marine applications [1], where glass-fiber-reinforced polymers are used as masts, submarine control surfaces, transmission shafts, and propellers. In engines for cars and aircrafts [2], thermosets are in contact with oil, which is used for cooling. Figure 3.1 shows the sorption of water in two amine-cured and anhydride-cured two-functional epoxy resins.

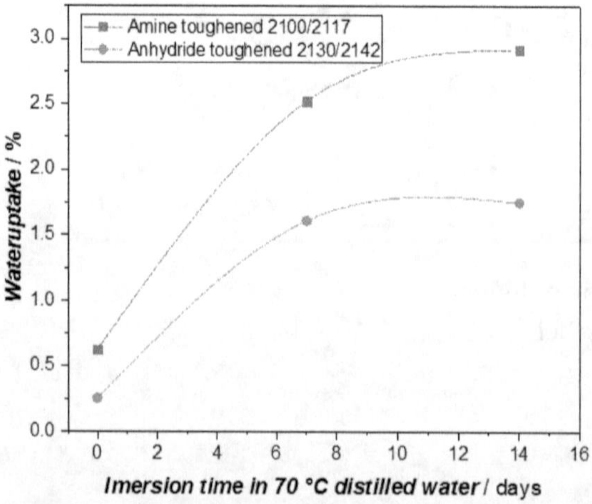

Figure 3.1 Water sorption in amine-cured and anhydride-cured two-functional epoxy resins

The materials are typically exposed to elevated temperatures. The gas diffusion comes into play for pressurized gas vessels [3, 4]. The recent developments in the automotive, trucking, and train industry has led to an increased demand of vessels to transport hydrogen and compressed natural gas.

Various methods exist to determine sorption and diffusion, and these can be divided on the basis of using mechanical, physical, gravimetrical, and dielectric methods. Depending on the application, the methods used for polymers and polymer composites should be chosen. The mechanical effect of sorption can be very effectively measured by dynamic-mechanical thermal analysis (DMTA) and by measuring the hardness. For the other mechanical tests as bending, creep, and tensile tests the geometry needs to be taken into account. Due to the defined geometries, the surface-to-volume ratio varies and influences the results, especially when the saturation is not reached yet [5]. By physical methods (such as calorimetric tests) and weighing the amount absorbed at a specific time the water intake was determined, while by contact angle measurements the polarity is measured, and can attribute to evaluating the results. Here also modified setups such as dynamic vapor sorption (DVS), as suggested by Preda, are recommended [6]. Chemical effects can be measured by near-infrared transmission (NIR) spectroscopy [7, 8], and dielectric spectroscopy has been used for epoxies by Hayward [9] and Pethrick [10].

3.2 Sorption of Water

3.2.1 Influence of Network Structure and Cure Kinetics

The diffusivity of liquids in thermosets strongly depends on the network structure, or more precisely on the free volume in the polymer network. The network structure is a result of chemical nature of the thermoset resin (polyester, epoxy, vinyl, or phenolic), the curing agent and thereby the curing reaction (polyaddition, polycondensation or polymerization), and the curing degree.

The chemical structure of the resin and hardener influences several factors, such as the network density and the polarity. For the **network density**, which is closely connected to the **free volume**, several attempts were made to define their influence on the absorption of liquids. Linde et al. used Jeffamine hardeners with varying molecular weight, which affected the network density to investigate their influence on the absorption of water [11]. Although a correlation could be found between crosslink density and water absorption for two of the curing agents used, samples with the curing agent with highest molecular weight showed the same water absorption as the sample with lower molecular weight. This clearly shows that crosslink density is not the only factor to take into account. In contrast, Frank et al. varied the crosslink density by two-, three-, and four-functional resin systems cured with diaminodiphenylsulfone (DDS) [12]. The approach is very systematic, as the chemical structure of the resulting polymer network is very similar. Their investigation clearly shows that the crosslink density is directly correlated to the uptake of liquids.

As the temperature rises above T_g for thermosets, their behavior significantly changes. The polymer gets significantly softer due to higher molecular moveability and reaches a rubbery state [13]. The thermal expansion coefficient increases rapidly, which directly correlates to the free volume, as shown by Greiner et al. [14]. The increased free volume then leads to an increase in water uptake.

As well as the network density and the free volume, the **polarity** of the polymer influences the absorption. Linde et al. showed that samples made from the amine hardener DDS showed significantly higher water uptake than other amine hardeners, despite a comparatively higher network density [11]. Also, varying the polarity by adding sidechains along the backbone influences the water absorption [15].

Furthermore, the **curing reaction** influences the absorption of liquids. In polycondensation, volatile products such as residuals lead to voids. As shown by Gellert et al., the voids play an important role in the water uptake of thermosets. They analyzed the behavior of glass-fiber-reinforced polyester, epoxy, vinyl, and phenolic resins. Phenolic resins showed a water update of 9% compared to 0.4% for the other resin systems, which led to a decrease in flexural strength by 25% for the corresponding fiber-reinforced composite, compared to 8% for the others [16]. In the case of phenolic resins, it has been shown that there is a higher water uptake and decrease in mechanical properties due to microporosity. Voids resulted from water in the resin and the polycondensation reaction [16].

The presence of aromatic rings during curing (in polyester resins) could be responsible for homopolymerization reactions, hydrophobic increase, volumetric shrinkage and incomplete resin polymerization [17, 18, 19], leading to a larger water sorption by these resins.

Concerning isophthalic and orthophthalic polyesters, their structural organization and good fiber-matrix affinity, coupled with lower steric hindrance, results in a more compact structure that prevents water diffusing through the bulk; the presence of low molecular weight species decreases durability in seawater [1].

This is further observed as vinyl ester polyesters have a better water sorption resistance when compared to isophthalic polyesters [20]. Indeed, a slower reaction of the vinyl resin cure leads to better organized linear macromolecular chains and a lower water diffusion coefficient but higher chemical stability, stiffness, and strength during its immersion in seawater than the isophthalic one [1]. Bunsell [21] related this behavior to a molecular rearrangement of the polymeric network when samples are thicker, which would slow down the water ingress via diffusion in the composite.

3.2.2 Effects on the Glass Transition Temperature and the Mechanical Properties

Water sorption in thermoset resins has also an influence on the macroscopic thermomechanical behavior of these materials. De'Nève and Shanahan [22] and Wright [23] showed that for bisphenol A diglycidyl ether *N,N*-dimethyldodecylamine (DGEBA-DDA) epoxy-amine resins, water sorption up to 4% leads to a drop in the elastic modulus measured by DMA as well as a drop in the glass transition temperature (T_g) from 115 °C to 90 °C, caused by a plasticization effect on such matrices. The same phenomenon was confirmed by Zhou and Lucas [24] for similar epoxy-amine resins. The mechanism behind this plasticization was attributed to water breaking the initial interchain Van der Waals force and hydrogen bonds resulting in the increase of chain segment mobility. Ellis and Karasz [25] demonstrated that drying and wetting epoxy-amine resins over time leads to the ageing of the materials, seen as a drop in their mechanical properties. Finally McKague, Reynolds, and Halkias [26] determined the same phenomena for polyester thermosets, i. e. water acting as a plasticizer for such resins, leading to a decrease in their mechanical properties.

3.2.3 Composites

Fiber-reinforced polymers are more susceptible to absorption of liquids than bulk polymers. A schematic representation of water diffusion in neat, filler-reinforced, and fiber-reinforced resins is shown in Figure 3.2.

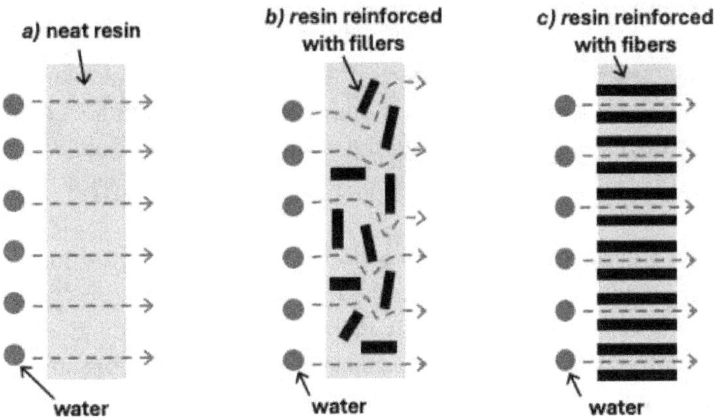

Figure 3.2 Water diffusion in neat, filler-reinforced, and fiber-reinforced resins

The interfaces between polymer and fiber are the weak point for the liquids to infiltrate the composite. Although the commonly used glass and carbon fibers themselves do not absorb liquids, they adsorb it through the fiber matrix interface [27, 28]. The mechanism of water sorption and its influence on composites was described in detail by Hammami et al. [17] for glass fibers as follows.

In the case of a relatively high fiber content, the fibers would be not fully impregnated by the resin. Thus, microcracks may appear at the fiber–matrix interface, which under the combined action of water diffusion and hydrogen ions, may accelerate the degradation process of the composite. For specimens immersed in seawater, uncured reactants were leached out, leaving the fibers unprotected. This resulted in the subsequent poor performance of such composites. Moreover, interlaminar shear strength decreased due to the poor bonding between the fiber and the matrix.

Specimens exposed to high temperatures and a fully saturated environment were subject to degradation caused by water diffusion through the matrix. Obviously, longer exposure periods would accelerate the initiation and propagation of the microcracks present at the surface, thus contributing to the diffusion process taking place within the matrix. Such an effect was confirmed by lower interlaminar shear strength values. Also, poor bonding between the fiber and the matrix may accelerate composite degradation, since it results in lower values of interlaminar shear strength. Specimens exposed to high temperatures and a completely dry environment show a decrease in their mechanical properties after a three-month period, although the variation (5%) was within the accepted range; while for a six-month period, values recorded for the mechanical properties were higher than those recorded with the virgin specimens. A possible explanation is that post-cure was fully achieved after the six-month period, thus providing high strength to the composite. Actually, visual observation of the specimens showed a discoloration of the composite: the original green resin changed to light yellow, which could be interpreted as resulting from further resin crosslinking.

Furthermore, as the interfaces are the weak part of the composite, the number of fiber cross-sections in contact with the surface and their length is substantial. In turn this is dependent on the fiber volume content and the orientation of the fibers [5]. As shown by Chilali at al., the diffusion rate and saturation weight increase with fiber orientation [5].

Concerning natural fibers, these are very sensitive to water adsorption [5, 29, 30, 31, 32]. Water molecules principally diffuse by capillarity through the micro-cracks present at the fiber–matrix interface. This significantly increases water diffusion kinetics within the composite and accordingly promotes more swelling of the fiber and a higher saturation weight compared to the first diffusion mode [5].

Boukhoulda et al. [33] studied the effect of fiber orientation while Neumann and Marom [34] studied the moisture absorption of carbon-epoxy composites of different fiber contents and fiber orientations under different mechanical stress levels. They found that there is no significant change in diffusion parameters when varying fiber orientation.

Finally, the residual mechanical properties for an immersed polyester/glass laminate were related to the absorbed water content [35, 36, 37]. It was observed that when the amount of water in the laminates increased, a reduction in the mechanical behavior was observed.

3.3 Oil Sorption

Oil uptake can be considered negligible for many thermoset resins, because their hydrophilic, polar nature prevents easy diffusion of often hydrophobic and non-polar oils such as bearing and transmission oils. Applications where contact of these fluids with thermosets and composites can take place can be found in automotive (gear box, electronic encapsulation, under-the-hood applications) or aviation (Skydrol in hydraulic systems). The variety and different composition of oils, lubricants, and composites makes general assumptions challenging. For example, the DOT 4 brake fluid consists of borate ester and glycol ethers, the hydraulic fluid Skydrol is based on phosphate esters, and for transmission fluids hydrocarbon lubricants are typically used.

Many oils are basic in nature to neutralize generated acids and prevent parts from corrosion.

3.3.1 Neat Resin

Neat resins and their oil uptake have rarely been investigated in the literature. Reuscher et al. [38] used an anhydride-cured epoxy resin with hexahydrophthalic anhydride and an imidazole-based hardener (DGEBA–HHPA–EMI with 100:83:1) and investigated the influence of ATF oil for 150, 165, and 180 °C for up to 1000 h. Aging was carried out above the initial T_g (142 °C) of the epoxy resin, which precludes future load-bearing applications. Still, despite being in the rubbery state, no mass uptake was visible, but a slight mass decrease of 0.5–1 wt% occurred, which can be linked to thermal degradation.

Braun et al. [39] investigated the influence of several industrial liquids on highly filled epoxy encapsulants at 120–160 °C. No weight uptake measurements were carried out, but the influence on CTE fracture toughness tests, T_g and modulus was investigated. Varley et al. [40] investigated the influence of Skydrol and toughener content on an epoxy resin, and no uptake was measurable.

3.3.2 Influence of Temperature

Most experimental investigations about the absorption of liquids remain limited to rather low temperatures. For marine applications, the temperatures are in accordance with the application, but especially in automotive or aviation applications where liquids such as hydraulic oils are used as a cooling medium, the thermoset parts are subjected to much higher temperatures. Service temperatures of up to 120 °C and a short-term temperature of up to 180 °C are common. Nevertheless, investigations with hydraulic oil were conducted at room temperature [40], 60 °C [37] or at the most 70 °C [3]. Markatos et al. showed that the effect of the hydraulic oil Skydrol was the same as the effect of moisture, with the fracture toughness decreasing by 27% in both instances [2]. The effect of moisture and oil was significantly less critical than thermal degradation, which led to a decrease of 70%. Figure 3.3 shows a schematic diagram of Skydrol sorption and diffusion in a neat epoxy resin. Skydrol sorption is usually comparable to that of water.

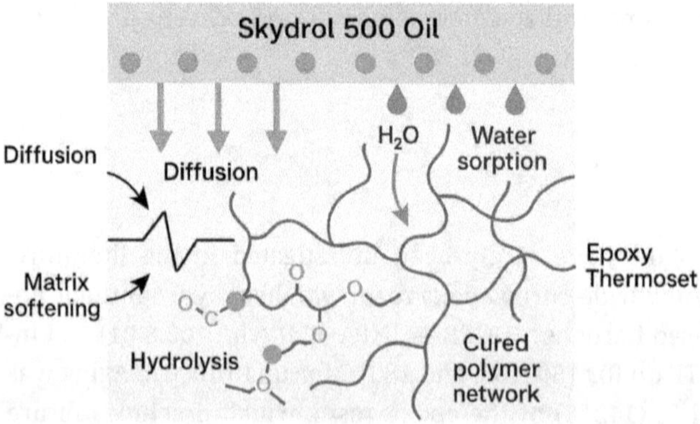

Figure 3.3 Schematic representation of Skydrol sorption and diffusion in neat epoxy resin, usually accompanied by water sorption

Varley et al. also found that water and Skydrol have similar effects on the thermomechanical properties of the used resins [40]. The effect on T_g is strongly dependent on the initial T_g of the system. The higher the difference between initial T_g and the testing temperature, the lower the effect of the immersion in liquids. They showed that the T_g for the used epoxy system dropped slightly from 180 °C to 175 °C after immersion for 120 days at room temperature in Skydrol [40].

3.3.3 Composites

Amaro et al. subjected glass fiber/epoxy composites to oil immersion tests at room temperature, using an universal multi-grade engine oil (15W-40) and an extra high performance hydraulic brake fluid (DOT 4) [41]. In order to study the effects of oil absorption, flexural and impact strength tests were carried out. For the 15W-40 a slight increase in weight with 0.38% after 45 days is visible, whereas for DOT 4 a decrease of 2.16% after the same period was observed.

Oscana et al. immersed epoxy adhesive joints in motor oil (Castrol Power 1 Racing 4T 10W-50) at 80 °C in a climatic chamber for periods of 1, 2, 4, 8, 16, 32, 52, and 128 h. Shear strength measurements were carried out, but no weight uptake of the epoxy resin was measured.

Boeing researchers [42] conducted a study with JP-4 fluid, Skydrol and fluid–water mixtures on ±45° carbon fiber/epoxy and aramide fiber/epoxy laminates for up to 5 years at ambient conditions. No significant influence was visible on the tensile properties for JP-4 fluid and Skydrol. In contrast to these findings for aramide fiber/epoxy, Sala et al. carried out static tensile and compression tests and dynamic compressive Woehler tests on epoxy/aramide fiber composites aged for 30 days at 60 °C in Skydrol [37]. The tensile modulus dropped to 89.6% and the compressive modulus to 82.6% of their initial values. The tensile strength decreased to 78.4% and the compressive strength to 60.4%. The bigger impact on the compressive properties can be explained with the matrix sensitivity of this test. Markatos et al. mixed Skydrol with water (50:50) to investigate the influence of the formed phosphorus acid on the epoxy-based HexPly® M32/T700 composite for 4 weeks at 70 °C [2]. A weight increase of 1.5% was detected and there was a decrease of 27% in interlaminar fracture energy G_{IC}. However, the weight increase can be linked mainly to water sorption.

References

[1] Visco, A. M.; Campo, N.; Cianciafara, P. Comparison of seawater absorption properties of thermoset resins based composites. *Composites Part A: Applied Science and Manufacturing* 2011, 42, 123–130

[2] Markatos, D. N.; Tserpes, K. I.; Rau, E.; Markus, S.; Ehrhart, B.; Pantelakis, S. The effects of manufacturing-induced and in-service related bonding quality reduction on the mode-I fracture toughness of composite bonded joints for aeronautical use. *Composites Part B: Engineering* 2013, 45, 556–564

[3] Michaels, A. S.; Vieth, W. R.; Barrie, J. A. Diffusion of gases in polyethylene terephthalate. *Journal of Applied Physics* 1963, 34, 13–20

[4] Meares, P. The diffusion of gases through polyvinyl acetate. *Journal of the American Chemical Society* 1954, 76, 3415–3422

[5] Chilali, A.; Assarar, M.; Zouari, W.; Kebir, H.; Ayad, R. Effect of geometric dimensions and fibre orientation on 3D moisture diffusion in flax fibre reinforced thermoplastic and thermosetting composites. *Composites Part A: Applied Science and Manufacturing* 2017, 95, 75–86

[6] Preda, F. M.; Alegría, A.; Bocahut, A.; Fillot, L. A.; Long, D. R.; Sotta, P. Investigation of water diffusion mechanisms in relation to polymer relaxations in polyamides. *Macromolecules* 2015, 48, 5730–5741

[7] Camacho, W.; Hedenqvist, M. S.; Karlsson, S. Near infrared (NIR) spectroscopy compared with thermogravimetric analysis as a tool for on-line prediction of water diffusion in polyamide 6,6. *Polymer International* 2002, 51, 1366–1370

[8] Musto, P.; Ragosta, G.; Mascia, L. Vibrational spectroscopy evidence for the dual nature of water sorbed into epoxy resins. *Chemistry of Materials* 2000, 12, 1331–1341

[9] Hayward, D.; Hollins, E.; Johncock, P.; McEwan, I.; Pethrick, R. A.; Pollock, E. A. The cure and diffusion of water in halogen containing epoxy/amine thermosets. *Polymer* 1997, 38, 1151–1168

[10] Pethrick, R. A.; Hollins, E. A.; McEwan, L.; Pollock, A.; Hayward, D.; Johncock, P. Effect of cure temperature on the structure and water absorption of epoxy/amine thermosets. *Polymer International* 1996, 39, 275–288

[11] Linde, E.; Giron, N. H.; Celina, M. C. Water diffusion with temperature enabling predictions for sorption and transport behavior in thermoset materials. *Polymer* 2018, 153, 653–667

[12] Frank, K.; Childers, C.; Dutta, D.; Gidley, D.; Jackson. M.; Ward, S.; Maskell, R.; Wiggins, J. Fluid uptake behavior of multifunctional epoxy blends. *Polymer* 2013, 54, 403–410

[13] Bard, S.; Demleitner, M.; Weber, R.; Zeiler, R.; Altstädt, V. Effect of curing agent on the compressive behavior at elevated test temperature of carbon fiber-reinforced epoxy composites. *Polymers* 2019, 11, 943

[14] Greiner, R.; Schwarzl, F. R. Thermal contraction and volume relaxation of amorphous polymers. *Rheologica Acta* 1984, 23, 378–395

[15] Ding, Y.; M. Liu; Li, S.; Zhang, S.; Zhou, W. F.; Wang, B. Contributions of the side groups to the characteristics of water absorption in cured epoxy resins. *Macromolecular Chemistry and Physics* 2001, 202, 2681–2685

[16] Gellert, E. P.; Turley, D. M. Seawater immersion ageing of glass-fibre reinforced polymer laminates for marine applications. *Composites Part A: Applied Science and Manufacturing* 1999, 30, 1259–1265

[17] Hammami, A.; Al-Ghuilani N. Durability and environmental degradation of glass-vinylester composites. *Polymer Composites* 2004, 25, 609–616

[18] Karbhari, V. M.; Wang, Q. Multi-frequency dynamic mechanical thermal analysis of moisture uptake in e-glass/vinylester composites. *Composites Part B: Engineering* 2004, 35, 299–304

[19] Karbhari, V. M. Dynamic mechanical analysis of the effect of water on e-glass-vinylester composites. *Journal of Reinforced Plastics and Composites* 2006, 25, 631–644

[20] Kootsookos, A.; Mouritz, A. P. Seawater durability of glass-and carbon-polymer composites. *Composites Science and Technology*, 2004, 64, 1503–1511

[21] Dewimille, B.; Bunsell, A. R. Accelerated ageing of a glass fibre-reinforced epoxy resin in water. *Composites* 1983, 14, 35–40

[22] De'Nève, B.; Shanahan, M. E. R. Water absorption by an epoxy resin and its effect on the mechanical properties and infra-red spectra. *Polymer* 1993, 34, 5099–5105

[23] Wright, W. W. The effect of diffusion of water into epoxy resins and their carbon-fibre reinforced composites. *Composites* 1981, 12, 201–205

[24] Zhou, J.; Lucas, J. P. Hygrothermal effects of epoxy resin. Part II: Variations of glass transition temperature. *Polymer* 1999, 40, 5513–5522.

[25] Ellis, T. S.; Karasz, F. E. Interaction of epoxy resins with water: The depression of glass transition temperature. *Polymer* 1984, 25, 664–669

[26] McKague, E. L.; Reynolds, J. D.; Halkias, J. E. Swelling and glass transition relations for epoxy matrix material in humid environments. *Journal of Applied Polymer Science* 1978, 22, 1643–1654

[27] Schutte, C. L. Environmental durability of glass-fiber composites. *Materials Science and Engineering R* 1994, 13, 265–323

[28] Kawagoe, M.; Doi, Y.; Fuwa, N.; Yasuda, T.; Takata, K. Effects of absorbed water on the interfacial fracture between two layers of unsaturated polyester and glass. *Journal of Materials Science* 2001, 36, 5161–5167

[29] Mokhothu, T. H.; John, M. J. Review on hygroscopic aging of cellulose fibres and their biocomposites. *Carbohydrate Polymers* 2015, 131, 337–354

[30] Le Duigou, A.; Davies, P.; Baley, C. Exploring durability of interfaces in flax fibre/epoxy micro-composites. *Composites Part A: Applied Science and Manufacturing* 2013, 48, 121–128

[31] Bessadok, A.; Marais, S.; Gouanvé, F.; Colasse, L.; Zimmerlin, I.; Roudesli, S.; Métayer, M. Effect of chemical treatments of alfa (*Stipa tenacissima*) fibres on water-sorption properties. *Composites Science and Technology* 2007, 67, 685–697

[32] Gautier, L.; Mortaigne, B.; Bellenger, V.; Verdu, J. Osmotic cracking nucleation in hydrothermal-aged polyester matrix. *Polymer* 2000, 41, 2481–2490

[33] Boukhoulda, B. F.; Adda-Bedia, E.; Madani, K. The effect of fiber orientation angle in composite materials on moisture absorption and material degradation after hygrothermal ageing. *Composite Structures* 2006, 74, 406–418

[34] Neumann, S.; Marom G. Prediction of moisture diffusion parameters in composite materials under stress. *Journal of Composite Materials* 1987, 21, 68–80

[35] Pritchard, G.; Speake, S. D. The use of water absorption kinetic data to predict laminate property changes. *Composites* 1987, 18, 227–232

[36] Lee, M. C.; Peppas, N. A. Models of moisture transport and moisture-induced stresses in epoxy composites. *Journal of Composite Materials* 1993, 27, 1146–1171

[37] Sala, G. Impact behaviour of heat-resistant toughened composites. *Composites Part B: Engineering* 2000, 31, 161–173

[38] Reuscher, V.; Haag, S.; Patzelt, G.; Mayer, B. Aging of anhydride-hardened epoxies in lubricants at elevated temperatures. *Journal of Applied Polymer Science* 2017, 134, 44877

[39] Adrian, J.; Braun, D.; Rode, K.; Pasch, H. New analytical methods for epoxy resins, 1. Liquid chromatography at the critical point of adsorption and MALDI-TOF mass spectrometry. *Die Angewandte Makromolekulare Chemie* 1999, 267, 73–81

[40] Varley, R. J.; Tian, W. Toughening of an epoxy anhydride resin system using an epoxidized hyperbranched polymer. *Polymer International* 2004, 53, 69–77

[41] Amaro, A. M.; Reis, P. N. B.; Neto, M. A.; Louro, C. Effect of different commercial oils on mechanical properties of composite materials. *Composite Structures* 2014, 118, 1–8

[42] [Anon], Contracts, *Aircraft Engineering and Aerospace Technology* 1986, 58, 31–33

4
Permeability of Oxygen through Polymers

Emmanuel Richaud and Alain Guinault

This chapter describes gas transfer phenomena within classical polymers – properties that are of particular interest in the case of thermoplastics used in the food packaging industry. Some structure–property relationships are described, together with some experimental methods for measurement. This chapter also focuses on the impact of oxygen diffusion on polymer degradation and durabilty, along with the phenomenon of diffusion-limited oxidation. Experimental methods for estimating the thickness of the oxidized layer are presented, together with some theoretical approaches aimed at predicting this phenomenon.

4.1 Theoretical Background

4.1.1 Permeability

Permeability describes the capability of a polymer to be infused by a liquid or a gas. In actual fact, there are two phenomena:

1. Where migration occurs mainly via micro-cracks and/or porosities
2. Where the polymer does not contain such defects, and which are the focus of this chapter. Here, the driving force is a pressure gradient, provoking three successive phenomena:
   - Absorption of the penetrant on the upstream surface
   - Dissolution and diffusion of the penetrant in the polymer
   - Desorption of the penetrant on downstream surface.

The time to reach equilibrium at the interface can be considered to be very short. This means that, from a kinetic point of view, the permeability can be approximated by the phenomena of dissolution and diffusion. The gas flux passing through a polymer can hence be expressed as:

$$F = -D \cdot A \cdot \frac{dc}{dz} \tag{4.1}$$

Where:

F = gas (or liquid) flow (mol s^{-1})

A = surface area (m^2)

D = diffusion coefficient ($m^2\ s^{-1}$)

c = gas concentration in the polymer (mol m^{-3})

z = depth in the diffusion direction (m)

When the two external surfaces of samples are at equilibrium (the gas concentrations being equal to c_1 and c_2), this equation can be simplified to:

$$F = -D \cdot A \cdot \frac{c_2 - c_1}{e} \tag{4.2}$$

Where:

e = total thickness of the sample

As will be seen later, the surface concentration of simple gases or small molecules can be first approximated from the external pressure (or activity) by Henry's law:

$$c = SP \tag{4.3}$$

Where:

S = solubility coefficient (mol m^{-3} Pa^{-1})

C = concentration of dissolved molecules in polymer (mol m^{-3})

P = pressure (Pa)

Combining Equation 4.2 and Equation 4.3 leads to:

$$F = D \cdot S \cdot A \cdot \frac{P_1 - P_2}{e} \tag{4.4}$$

For gases obeying Henry's law, the product of diffusivity and solubility is also named the permeability (Pe):

$$\text{Pe} = D \cdot S = \frac{ne}{A \cdot \Delta t \cdot \Delta P} \tag{4.5}$$

Where:

n = amount of substance (mol)

Δt = time taken to diffuse through the sample (s)

According to Equation 4.5, Pe is in mol m^{-1} s^{-1} Pa^{-1}, but since S can also be expressed in cm$^{3}_{STP}$ cm$^{-3}_{polymer}$ Pa^{-1}, permeability is expressed in the literature in several other units. For example, the "barrer" corresponds to 10^{-10} cm$^{3}_{STP}$ cm cm^{-2} s^{-1} cm$_{Hg}^{-1}$.

The permeability coefficient Pe is an intrinsic property of a polymer. It must not be confused with permeance, which is the ratio of permeability over thickness:

$$\text{Permeance} = \frac{\text{Pe}}{e} = \frac{n}{A \cdot t \cdot \Delta P} \tag{4.6}$$

Accordingly, for a given polymer, a lower permeance value means a better performance.

Another value can be found in literature: the oxygen transmision rate (OTR):

$$\text{GTR} = \frac{n}{A \cdot t} = \frac{\text{Pe} \cdot \Delta P}{e} \tag{4.7}$$

This value is the result given directly by the classical devices for measuring permeability, and does not take into account the thickness of the material and the partial pressure of the gas applied to the surface. It is used classically by industry to ensure film processing quality.

4.1.2 Diffusion

Diffusion expresses the kinetic aspect of permeation. It basically corresponds to the effect of penetrant size, polymer free volume (with macromolecular motions favoring this), and can be affected by tortuosity due to fillers or crystallites. Two mechanisms of gas diffusion can be distinguished and will be presented in this section.

4.1.2.1 Fickian Diffusion

This process occurs when the characteristic diffusion time is much higher than the typical relaxation time of polymer, which means that the diffusion rate is lower than the polymer relaxation rate. In other words, it is frequently observed in polymers well above their glass transition temperature [1]:

$$J = -D \cdot \frac{\mathrm{d}c}{\mathrm{d}x} \tag{4.8}$$

$$\frac{\mathrm{d}c}{\mathrm{d}t} = D \cdot \frac{\mathrm{d}^2 c}{\mathrm{d}x^2} \tag{4.9}$$

Where:

J = flow of penetrant through the polymer film (mol m^{-2} s^{-1})

D = diffusion coefficient of penetrant (m^{2} s^{-1})

c = concentration (mol m^{-3}) at position x and the time t

4.1.2.2 Non-Fickian Diffusion and Anormal Diffusion

These diffusion mechanisms are observed when the diffusion rate of penetrant is comparable to the relaxation rate of relaxation (in the case of anormal diffusion) or faster (in the case of non-Fickian transport). In that case, the diffusion coefficients changes with time according to the following equation:

$$\frac{\partial D(c,t)}{\partial t} = \frac{\partial D}{\partial c} \cdot \frac{\partial c}{\partial t} + \alpha \cdot (D_e - D) \quad (4.10)$$

Where:

D_e = diffusion coefficient at equilibrium in the fully relaxed matrix ($m^2\ s^{-1}$)

α = relaxation rate coefficient (s^{-1})

4.1.3 Solubility

Solubility describes the mechanims of "dispersion" of the gas in the polymer matrix. This mechanism depends on the experimental conditions (temperature, pressure) and the polymer nature. The solubility coefficient S (expressed either in mol $m^{-3}\ Pa^{-1}$ or in $cm^3_{STP}\ cm^{-3}\ atm^{-1}$) is a thermodynamic value representing the number of gas molecules dissolved in polymer at equilibrium, and is linked to polymer–gas interactions.

4.1.3.1 Henry's Law

Henry's equation (Equation 4.3) describes a linear sorption isotherm, and is commonly used for describing the sorption of simple gases in rubber [2].

4.1.3.2 Double-Mode Sorption Model

This model was first proposed by Meares [3]. It describes the case of glassy polymers with empty microholes (related to free volume). The dissolution of the penetrant in the latter is responsible for a supplementary term (c_{holes}) added to the term for the Henry's law dissolution mechanism (c_{Henry}) in the equation, giving the sorption isotherm [4, 5]:

$$c = c_{Henry} + c_{holes} \quad (4.11)$$

$$c = k_{Henry} \cdot p + \frac{k'_{holes} \cdot b \cdot p}{1 + (b \cdot p)} \quad (4.12)$$

Where:

c = solubility ($cm^3_{STP}\ cm^{-3}_{polymer}$)

k_{Henry} = Henry's law dissolution constant ($cm^3_{STP}\ cm^{-3}_{polymer}\ Pa^{-1}$)

k'_{holes} = hole saturation constant ($cm^3_{STP}\ cm^{-3}_{polymer}$)

b = hole affinity constant (Pa^{-1})

Formally, the second term of Equation 4.10 expresses a Langmuir dissolution term [6]. In essence, the holes responsible of this latter are formed in glassy materials. This observation can be illustrated in Figure 4.1, which depicts SO_2 sorption in polymers with an amorphous phase, specifically PE (with a rubbery phase), PC (with a glassy phase), and PA6 (which can be considered as an intermediate case between PE and PC).

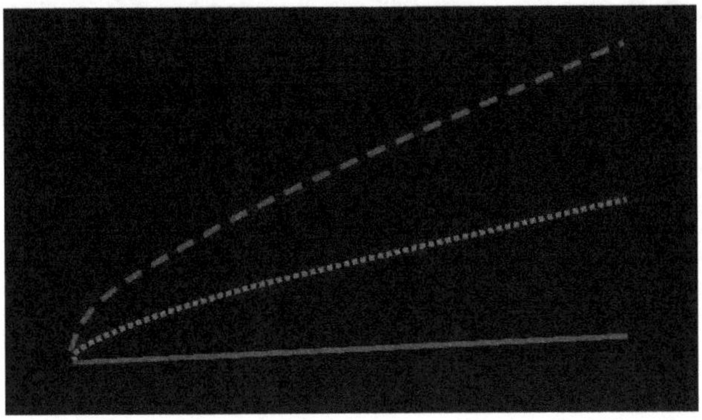

Figure 4.1
Sorption of SO_2 in rubbery or glassy polymers [7]

Other examples of double sorption have been (for example) observed for CH_4 in PS [8] and CO_2 in PC [9]. Lately, Haurat et al. [10] have also illustrated the "transition" from a double sorption mechanism to a Henry's law mechanism when passing through the glass transition for PMMA-based polymer blends.

4.1.4 Effect of Temperature

The solubility coefficient S (expressed either in mol m^{-3} Pa^{-1} or in cm$^{3}_{STP}$ cm^{-3} atm^{-1}) represents a thermodynamic equilibrium, and is given by a Van 't Hoff law:

$$S = S_0 \cdot \exp\left(-\frac{\Delta H_S}{R \cdot T}\right) \tag{4.13}$$

Since D is also given by an Arrhenius law,

$$D = D_0 \cdot \exp\left(-\frac{\Delta H_D}{R \cdot T}\right) \tag{4.14}$$

permeability can thus be given by an apparent Arrhenius law:

$$\mathrm{Pe} = \mathrm{Pe}_0 \cdot \exp\left(-\frac{E_P}{R \cdot T}\right) \tag{4.15}$$

Where:

Pe_0 (pre-exponential factor) = $D_0 \cdot S_0$

E_p (activation energy) = $\Delta H_S + E_D$

Some typical values are given in Table 4.1. E_D is generally of the order of 20–30 kJ mol^{-1}, and E_S seems to vary from values approximatively equal to 0 to values around $2 \times E_D$.

Table 4.1 Typical Permeation Parameters for Oxygen in Polymers at Room Temperature under Dry Atmosphere

Polymer	T (°C)	Pe (cm^3 cm cm^{-2} s^{-1} Pa^{-1})	D (cm^2 s^{-1})	E_D (kJ mol^{-1})	S (cm$^3_{STP}$ cm$^3_{polymer}$ Pa^{-1})	E_S (kJ mol^{-1})	Ref.
LDPE	RT	(1958–2082) × 10^{-16}	3.1×10^{-7}	—	8.4 × 10^{-7}	—	[11]
HDPE	RT	(625–832) × 10^{-16}	(1.9– 2.7)×10^{-7}	—	(2.9-5.2) × 10^{-7}	—	[11]
LLDPE	RT	1920 × 10^{-16}	(2-5)×10^{-7}	~35	(7.5-31) × 10^{-7}	—	[11, 12, 13]
PP	RT	—	0.87×10^{-7}	—	4.5 × 10^{-7}	—	[14]
EVA19	25	4.028 × 10^{13}	—	—	—	—	[14]
EVA33	25	7.98 × 10^{-13}	—	—	—	—	—
EVA50	25	6.156 × 10^{-13}	—	—	—	—	—
EVA70 (70% VA)	25	2.736 × 10^{-13}	—	—	—	—	—
PA 6I/6T	11.9	—	~3 × 10^{-10}	~35	3.5 × 10^{-6}	~-5	[15]
	22	—	~6 × 10^{-10}		3.0 × 10^{-6}		
	40.3	—	~11.5 × 10^{-10}		2.8 × 10^{-6}		
PA66	—	(8.99–9.08) × 10^{-16}	—	—	—	—	—
EVOH (32%Et)	—	7 × 10^{-18}	—	—	—	—	—
EVOH (44% Et)	—	4.4 × 10^{-17}	—	—	—	—	—
Epoxy 1	30	—	2.4 × 10^{-9}	~10	4 × 10^{-6}	~40	[16]
	40	—	2.8 × 10^{-9}	—	6.85 × 10^{-6}	—	—
Epoxy 2	10	—	1.5 × 10^{-9}	5	1.5 × 10^{-6}	40	[19]
	20	—	1.6 × 10^{-9}	—	2.1 × 10^{-6}	—	—
	30	—	1.6 × 10^{-9}	—	4.7 × 10^{-6}	—	—

Dealing now with the double sorption model, this is observed in polymers in the glassy state but not those in a rubbery state. It is hence not surprising to observe that sorption can obey the "double sorption model" at temperatures below T_g and that the curvature associated with this model disappears when the temperature increases. In the rubbery state, the sorption model eventually obeys Henry's law (Figure 4.2).

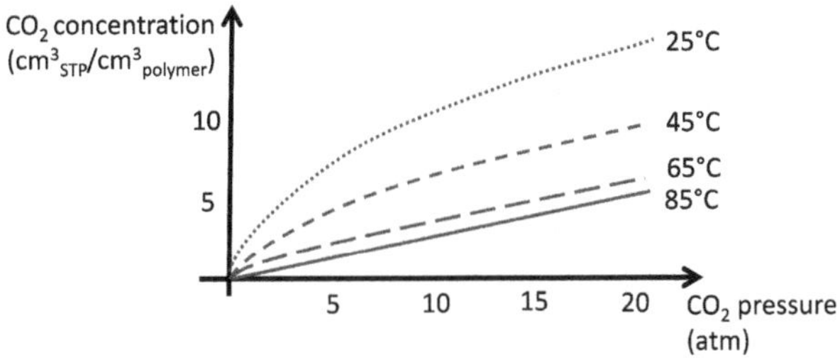

Figure 4.2 CO_2 sorption into PET at various temperatures (adapted from [17])

4.1.5 Selectivity

A polymer can be defined by its selectivity towards two distinct gaseous species. The selectivity can be defined as:

$$\alpha_{ij} = \frac{P_i}{P_j} = \frac{D_i}{D_j} \cdot \frac{S_i}{S_j} = \alpha_{ij}^{D} . \alpha_{ij}^{S} \tag{4.16}$$

4.2 Methods for Measuring Permeability and Related Values

Various methods can be used to measure permeability, diffusivity, and solubility. The classical principle of experimental devices is to measure the gas flux though a polymer film or sheet, but other shapes can be used by adjusting the experimental setup (bottles, tubes, etc.).

4.2.1 Measurement of Permeability

Conventionally, two methods are used for measuring the permeability coefficient: the isobaric (or equal pressure) method (ISO 15105-2:2003) and the differential pressure method (ISO 15105-1:2002), as depicted in Figure 4.3. Both methods consist of applying a flow of the desired gas to one face of the material (often used as a film or sheet) and measuring on the second face the number of moles or flow that passes through the material [18].

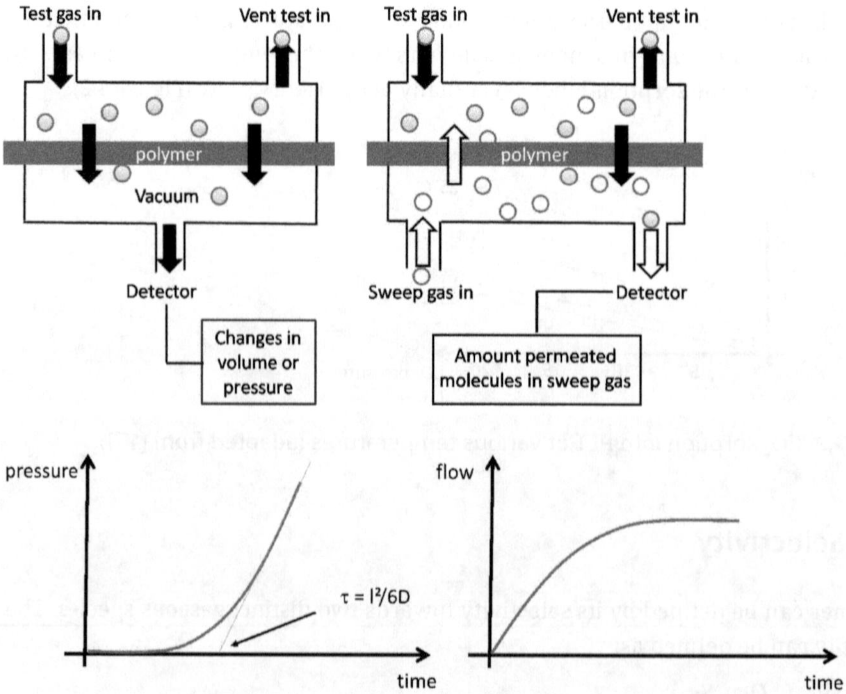

Figure 4.3 Top: Schematic showing gas permeability for (left) the differential pressure method and (right) the equal pressure method. Bottom: Corresponding changes of (left) the back-pressure and (right) the instantaneous gas flow, as a function of time

In the isobaric (or equal-pressure method), the permeance (sometimes called gas transfer coefficient or GTC) is given by:

$$GTC_{O_2} = \frac{C \cdot \varphi}{A \cdot P_0} \tag{4.17}$$

Where:

φ = flux of test gas ($cm^3\ s^{-1}$)

C = concentration of gas detected as a volume fraction (i. e. in $cm^3_{gas}\ cm^{-3}_{sweep\ gas}$)

P_0 = oxygen partial pressure in test gas (Pa)

A = film surface area (cm^2)

For the isobaric pressure, the gas flow is measured by gas chromatography–mass spectrometry for the case of gas mixtures [11, 19, 20], or (sometimes) using other sensors based on electrochemical methods [13] such as the reduction of oxygen:

$$O_2 + H_2O + 4e^- \rightarrow 4HO^-$$

$$4Ag \rightarrow 4Ag^+ + 4e^-$$

Concerning the differential pressure method [21], both faces of the material are placed under vacuum before the gas is applied on one face of the material and the pressure increase is measured on the back face. The permeability is derived from:

$$\mathrm{Pe} = \frac{e}{\Delta P \cdot A} \cdot \frac{V_\mathrm{c} \cdot V_\mathrm{m}}{R \cdot T} \left(\frac{\mathrm{d}P}{\mathrm{d}t}\right)_\mathrm{ss} \tag{4.18}$$

Where:

Pe = gas permeability coefficient (barrer = 1×10^{-10} $\mathrm{cm^3_{STP}\ cm\ cm^{-2}\ s\ cm_{Hg}}$)

V_c = downstream cell volume (cm^3)

V_m = volume of a gas mole at standard conditions (cm^3)

e = film thickness (cm)

A = effective membrane area (cm^2)

T = absolute temperature (K)

R = gas constant

ΔP = pressure difference between upstream and downstream (Pa)

$(\mathrm{d}p/\mathrm{d}t)_{ss}$ = steady-state increase of downstream pressure (Pa s^{-1})

As illustrated later, the apparent diffusion coefficient (D) is calculated from the time lag before reaching steady-state permeation.

Depending on the method of the measurement (instantaneous flow with a specific sensor, or integrated flow using a pressure increase measurement), the permeability coefficient can be calculated when the system reaches equilibrium, as explained in Section 4.1.

4.2.2 Diffusion Coefficient Measurement

The most popular method used for determining the diffusion coefficient is often named the timelag method. It is deduced from the integrated flow curve or the pressure curve depending on the method of permeability, as illustrated in Figure 4.3.

According to the mathematical equations developed by Cranck [22], the concentration of gas in the polymer matrix is given by:

$$c = c_1 + (c_2 - c_1)\frac{x}{l} + \frac{2}{\pi}\sum_{n=1}^{\infty} \frac{c_2 \cdot \cos(n \cdot \pi) - c_1}{n} \cdot \sin\left(\frac{n \cdot \pi \cdot x}{l}\right) \cdot \exp\left(-\frac{D \cdot n^2 \cdot \pi^2 \cdot t}{l^2}\right) \tag{4.19}$$

The maximal slope (i. e. the asymptotic behavior when *t* tends towards infinity) tends towards:

$$Q = \frac{D \cdot c_1}{e} \cdot \left(t - \frac{e^2}{6D}\right) \tag{4.20}$$

The intersect of the maximal slope and time axis (in Figure 4.3) is given by:

$$\tau = \frac{e^2}{6D} \tag{4.21}$$

The solubility coefficient *S* can thus be obtained as $c_1 = Q/D$.

4.2.3 Solubility Coefficient Measurement

Apart using permeability methods described above, the solubility of simple gases can be estimated by measurement ($P = D{\cdot}S$) for gases following Henry's law) or directly by several methods such as gravimetry or spectroscopy [23]. For instance, CO_2 can be identified from its specific stretching and bending mode as described in Table 4.2, and can be quantified (for example) using $\varepsilon = 0.25\ \mathrm{L\,mol^{-1}\,cm^{-1}}$ at $4950\ \mathrm{cm^{-1}}$; however, such weak molar absorptivity means that those methods are only sensitive for weight ratios around 10%.

Table 4.2 Spectroscopic Determination for Simple Gases in Polymers

	Wavenumber (cm^{-1})	Contribution
CO_2 [29]	2350	υ_{AS}
	3590	$2\delta + \upsilon_{AS}$
	3695	$\upsilon_S + \upsilon_{AS}$
	4850	$\upsilon_{AS} + 4\delta$
	4950	$\upsilon_S + \upsilon_{AS} + 2\delta$
	5030	$2\upsilon_S + \upsilon_{AS}$
SO_2 [24]	518	δ
	1151	υ_S
	1362	υ_{AS}
	2500	$\upsilon_S + \upsilon_{AS}$

4.3 Experimental Values and Structure–Property Relationships

4.3.1 Effect of Penetrant Size and Structure

Table 4.3 compares the solubility of simple gases in PDMS. It is observed that $S_{CO_2} > S_{CH_4} > S_{O_2} > S_{N_2}$. In the case of sorptions obeying Henry's law, a general law was proposed [25]:

$$\ln S = b \cdot T_c + a \tag{4.22}$$

Where:

T_c = critical temprature

Other correlations are proposed with the boiling temperature, or Lennard-Jones (or critical) temperature of chemicals in the case of their solubility in natural rubber [26].

Table 4.3 Sorption of Simple Gases in PDMS

Gas	T (°C)	S (cm^3_{STP} cm^{-3} atm^{-1})	D (10^{-5} cm^2 s^{-1})	Ref.
N_2	35	0.1	3	—
O_2	35	0.2	3.2	[27]
CH_4	35	0.42	2.5	—
CO_2	35	1.18	2.6	—

In principle, one has $D_{CO_2} < D_{N_2} < D_{O_2}$, as observed for example in PLA [28, 29] or PDMS [27]. This trend agrees well with the steric hindrance according to the Van der Waals diameters of the penetrant molecules, which are 42.67, 39.13, and 31.83 cm^3 mol^{-1} for CO_2, N_2, and O_2, respectively. However, contradictory trends have beem observed, suggesting that diffusivity is linked to the kinetic diameter of molecules [28].

4.3.2 Effect of Polymer Structure and Crystallinity

Gas diffusion of molecules through polymers can be predicted from a theory based on free volume:

$$D = A \cdot \exp\left(-\frac{B}{V_f}\right) \tag{4.23}$$

Where:

B = size of penetrant, which determines its motion

V_f = free volume content

V_f can be defined as the excess of volume compared to the volume actually occupied by atoms:

$$V_f = \frac{v - v_0}{v_0} \tag{4.24}$$

In linear polymers, molar mass is shown to be related to many physical properties of the polymer. This is the case for transport properties, as illustrated by several amorphous polymers in the glassy state [30], where a correlation is observed between the diffusivity of organic vapors and the number-average molecular weight:

$$\log D = \log D_\infty + a/M_n \tag{4.25}$$

Formally, this is very close to the Flory–Fox law for the glass transition [31]. This in in fact not surprising, since both laws express the increase of free volume due to chain ends (responsible for the T_g decrease and diffusivity increase).

As recalled in the previous paragraphs, the solubility of gases in glassy polymers can be depicted by a double sorption model. In that case, the solubility parameters given in Equation 4.12 could be calculated from an additive method [32]:

$$k_D = \sum_{i=1}^{C} \phi_i \cdot k_{D_i} \tag{4.26}$$

$$C'_H = \sum_{i=1}^{C} \phi_i \cdot C'_{H_i} \tag{4.27}$$

$$b = \sum_{i=1}^{C} \phi_i \cdot b_i \tag{4.28}$$

$$\phi_i = \frac{\text{Van der Waals volume of group } i}{\text{Van der Waals volume of polymer}} \tag{4.29}$$

It seems obvious that diffusivity must tends towards 0 in a 100% crystalline sample. Van Krevelen [26] has thus proposed the following relationship.

$$D = D_{am} \cdot (1 - x_C) \tag{4.30}$$

Where:

D_{am} = diffusion in a 100% amorphous sample

In an alternative theory, it is considered that diffusion is thermally activated, so that diffusivity can be written as:

$$D = \upsilon \cdot \exp\left(-\frac{\Delta G^{++}}{R \cdot T}\right) \tag{4.31}$$

Where:

ΔG^{++} = free energy for a semi-crystalline polymer, and the prefactor value is:

$$\upsilon = \frac{k_B \cdot T}{h} \cdot \lambda^2 \tag{4.32}$$

Where:

k_B = Boltzmann constant

h = Planck constant

λ = characteristic distance for a jump of the penetrant molecule

According to Compañ et al. [33], the activation energy is given by:

$$\Delta G^{++} = x_{am} \cdot \Delta G^{++}{}_{am} = (1 - x_C) \cdot \Delta G^{++}{}_{am} \tag{4.33}$$

Where:

$\Delta G^{++}{}_{am}$ = activation free energy for a 100% amorphous polymer

If $D^*(T)$ and $D(x_{am},T)$ respectively denote the diffusion coefficient in the amorphous phase, and in a semi-crystalline polymer with a crystalline ratio equal to x_C, the following relationship can be established:

$$\frac{D^*(T)}{D(x_{am}, T)} = \exp\left(-\frac{\Delta G^{++}{}_{am} - \Delta G^{++}}{R \cdot T}\right) = \exp\left(-\frac{\Delta G^{++}{}_{am}}{R \cdot T} \cdot x_C\right) \tag{4.34}$$

The expression proposed by Compañ et al. [33] is consistent with the one proposed by Michael and Bixler [34]:

$$\frac{D^*(T)}{D(x_{am}, T)} = \beta \cdot \tau \tag{4.35}$$

Where:

β = immobilization of amorphous phase by crystalline phase

τ = tortuosity factor due to the hindrance of gas diffusion by crystallites

These can be given by:

$$\beta = \exp\left(-\frac{\Delta H^{++}{}_{am} - \Delta H^{++}}{R \cdot T}\right) \quad (4.36)$$

$$\tau = \exp\left(\frac{\Delta S^{++}{}_{am} - \Delta S^{++}}{R}\right) \quad (4.37)$$

This means that:

$$\Delta S^{++}{}_{am} - \Delta S^{++} = R \cdot \ln(\tau) \quad (4.38)$$

Where:

τ = number of conformation of obstacles to diffusion, and is equal to 1 for 100% amorphous polymers

The effect of crystallinity on barrier properties can be triggered by orientation effects combined with the process. This is illustrated in Figure 4.4 as a function of the draw ratio, [35], where LLDPE permeability is compared for unoriented films, and for films stretched in a parallel or transverse direction compared to extrusion. It is clearly observed that orientation has a very favorable role on permeability, in particular in the parallel direction.

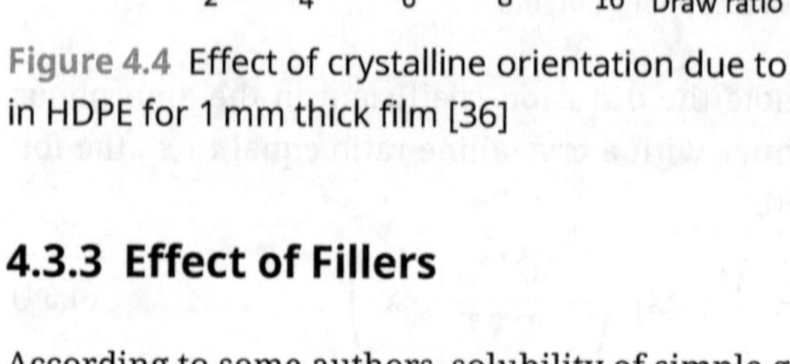

Figure 4.4 Effect of crystalline orientation due to the draw ratio on O_2 permeability in HDPE for 1 mm thick film [36]

4.3.3 Effect of Fillers

According to some authors, solubility of simple gases should obey a mixture law:

$$S = (1 - \varphi) \cdot S_{polymer} + \varphi \cdot S_{filler} \quad (4.39)$$

Where:

φ = volume fraction of filler

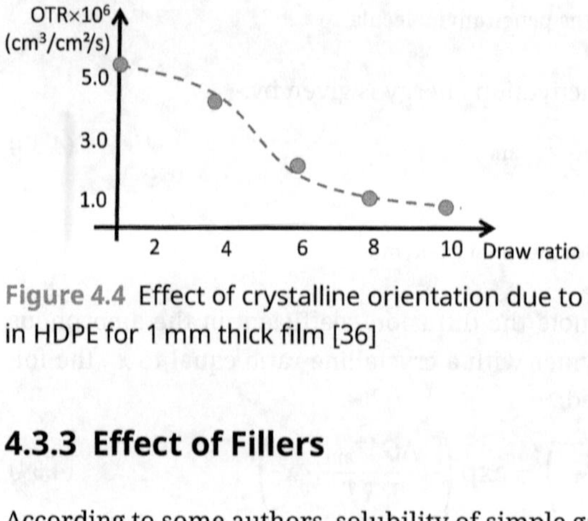

Where:

In case they are not permeable to oxygen, fillers have a very similar influence to crystallinity, since they increase tortuosity. However, Equation 4.30 seems valid up to a limited nanofiller weight ratio, above which a distortion in the mixture law can be observed (see for example [37]). The most suitable expression is [38]:

$$\frac{D}{D_0} = \frac{1}{\tau} \tag{4.40}$$

Where:

τ = toruosity

It is expected that fillers decrease overall permeability. In general, they slow down diffusion meanwhile their effect on solubility remains moderate [27]. Under the assumption that $S_{\text{filler}} \ll S_{\text{polymer}}$, Equation 4.32 and Equation 4.33 can be combined, which gives:

$$\frac{\text{Pe}}{\text{Pe}_0} = \frac{1-\varphi}{\tau} \tag{4.41}$$

The tortuosity can be expressed from the volume fraction (φ) and aspect ratio of fillers (α) of the fillers, as illustrated for example in [39, 40]. General expressions for τ are given in Table 4.4.

Table 4.4 Effect of Fillers on Permeability

Dispersion type	Alignment	Tortuosity factor, τ	Schematic
Random, dilute	Random	$1+\frac{\alpha}{6}\varphi$	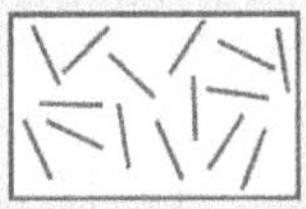
Random, dilute	Aligned	$1+\frac{\alpha}{2}\varphi$	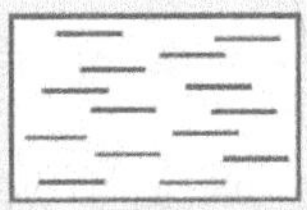
Random, semi-dilute	Aligned	$\left(1+\frac{\alpha}{3}\varphi\right)^2$	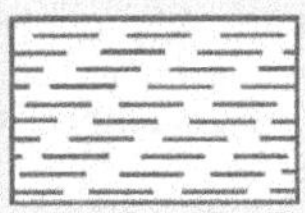

4.3.4 Blends and Multi-Layered Polymers

Let us start by comparing the barrier properties of several thermoplastics to oxygen and water. It is easy to see that polymers, being a good barrier to oxygen, can be very permeable to water and vice versa. Typically:

- Polyethylene (in particular HDPE) and PP are highly permeable to oxygen but display good barrier properties to water
- PA6 or EVOH display good barrier properties to oxygen (in particular with higher vinyl alcohol groups) but water easily passes through.

Finally, apart from PET and PVDC, which present a reasonable compromise, achieving a polymeric package that has good barrier properties for several kinds of chemicals (for example oxygen and water) requires the use of a multilayered structure [28, 41], such as PE/EVOH/PE.

The permeability of multilayered polymers can thus be expressed as:

$$Pe_T = \frac{\sum x_i}{\sum \frac{x_i}{Pe_i}} \tag{4.42}$$

Where:

Pe_i = permeability of each layer

x_i = thickness of each layer

Blends or multilayers can also be used, with the advantage of simpler process. Their permeability can be predicted using:

$$Pe_{blend} = Pe_c \left[1 + \frac{3\phi_d}{\frac{Pe_d + 2Pe_c}{Pe_d - Pe_c} - \phi_d} \right] \tag{4.43}$$

Where:

Pe_{blend} = blend permeability

Pe_c = permeability of the continuous phase

Pe_d = permeability of the dispersed phase

ϕ_d = volume fraction of the dispersed phase

However, a comparison of Equation 4.41 and Equation 4.39 shows that the "multilayer strategy" is more efficient for decreasing permeability than simple blending. This is in particular true in the case where the dispersed phase has a lamellar structure.

4.3.5 Effect of Water Concentration

It seems clear that modifying the external water concentration may change the parameters for gas permeability, as in the case of polyamides [42] (Figure 4.5):

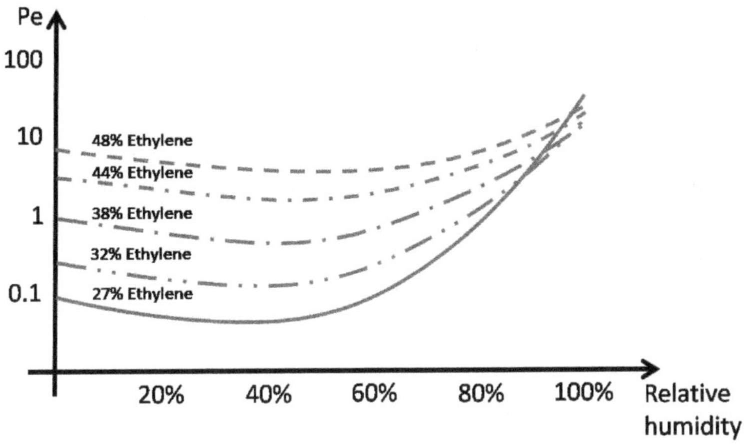

Figure 4.5 Permeability (in cm^3 20 µm m^{-2} day^{-1} atm^{-1}) for EVOH with various ethylene contents [43]

4.4 The Role of Oxygen Diffusion in Polymer Degradation

4.4.1 Brief Overview of Diffusion-Limited Oxidation

It has been established for a long time that most polymers can undergo oxidation mechanisms. The in-chain reactions between polymer and oxygen lead to chain scission or crosslinking, changing the polymer's properties and resulting in failure. Two "extreme" cases must be distinguished:

1. Thin films, where oxidation can be considered to be homogeneous. Here, it is considered that failure results when a critical state is reached (for example a critical value of molar mass in the case of a thermoplastic polymer)
2. Thick samples, where failure results from the balance between oxidized (degraded) layers and bulky (unoxidized) ones.

To better understand the meaning of this layer degradation and help us to predict when it happens, let us recall that the oxygen consumption rate is given by a balanced equation expressing the coupling between diffusion and reaction:

$$\frac{\partial[O_2]}{\partial t} = D_{O_2} \cdot \frac{\partial[O_2]}{\partial x^2} - r_{ox} \tag{4.44}$$

Where:

r_{ox} = quantity of oxygen reacting in an elementary layer

At a steady state, an equilibrium is reached and the thickness of the oxidized layer plateaus at a value denoted by the "thickness of the oxidized layer" (TOL), approximated by [44]:

$$\mathrm{TOL}^2 = D_{O_2}[O_2]/r_{ox} \tag{4.45}$$

There are many papers describing changes of oxidation levels or mechanical properties within the depth of aged polymer samples [45, 46, 47, 48, 49], with the two being tightly correlated (Figure 4.6):

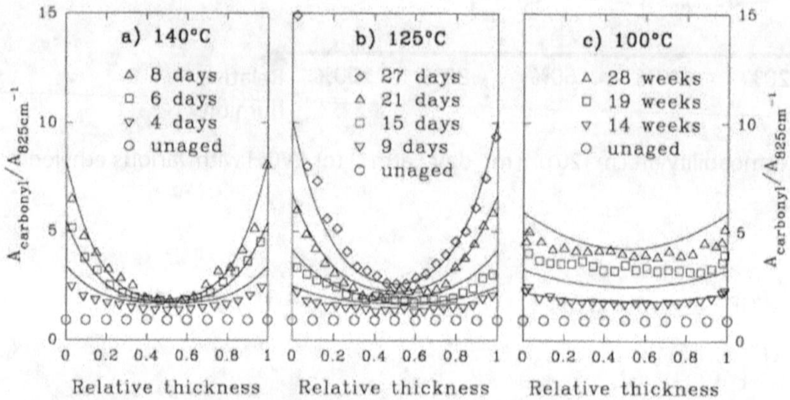

Figure 4.6 Variation of oxidation-tracking carbonyls with relative thickness, for a rubber thermally oxidized at various temperatures for different times [50]

Those results confirm that oxidation only occurs in a superficial layer. Another method for characterizing this latter is to monitor mass loss for samples with various thicknesses (Figure 4.7):

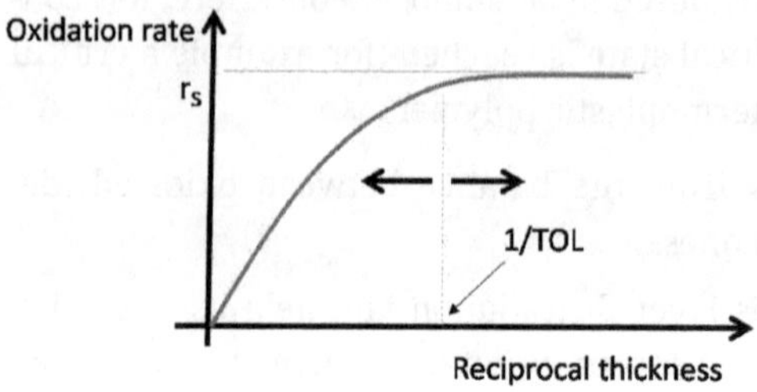

Figure 4.7 Schematic showing apparent oxidation (estimated for example from mass loss) vs reciprocal thickness (adapted from [48])

It can be observed that there is:

- A quasi-linear increase of oxidation rate in the domain of high thickness
- A plateau where oxidation rate becomes constant when thickness gets lower than a critical value (denoted by TOL). This latter seems very close to the value of the thickness of the superficial layer where degradation can be observed visually [48].

Based on Equation 4.45, it can be seen that changes of TOL with temperature can be described by an apparent Arrhenius law with:

$$E_{TOL} = E_D - E_S - E_{OX} \quad (4.46)$$

Where:

E_{TOL} = activation energy for the thickness of oxidized layers

E_D = activation energy for oxygen diffusivity

E_S = activation energy for oxygen solubility

E_{OX} = activation energy for oxidation rate

E_{OX} is usually much higher than 50 kJ mol^{-1}, so that E_{TOL} is negative. This means that TOL is expected to decrease when increasing ageing temperature (as illustrated for example in Figure 4.6). From a practical point of view, it complicates the direct realization of "representative accelerated ageing", since the latter usually involve enhanced temperatures, meaning that the damaged layer may be significantly thinner in accelerated ageing than in the "real" one. The issue of TOL prediction is thus crucial for researchers and engineers in charge of predicting the durability of polymers and composite materials.

4.4.2 Prediction of the Thickness of Oxidized Layer

In the case of thermal oxidation, there are several methods for predicting the apparent rate for oxygen consumption (r_{ox}). One of them is to propose a mechanistic scheme involving the main reactive species (substrate radical (P$^\bullet$), peroxide radical (POO$^\bullet$), substrate (PH), hydroperoxide (POOH), oxygen, etc.) and derive a differential system. This can then be resolved so as to: (i) compare with experimental observations of the concentration in reactive species; (ii) validate the model; (iii) extrapolate the rate constant values that "rule" the reactivity; and (iv) predict the oxidation rate. The following mechanisms can be used in an initial approach:

$\delta POOH \rightarrow \alpha P^{\cdot} + \beta POO^{\cdot} + carbonyls$	k_1
$P^{\cdot} + O_2 \rightarrow POO^{\cdot}$	k_2
$POO^{\cdot} + PH \rightarrow POOH + P^{\cdot}$	k_3
$P^{\cdot} + P^{\cdot} \rightarrow$ stable products	k_4
$P^{\cdot} + POO^{\cdot} \rightarrow$ stable products	k_5
$POO^{\cdot} + POO^{\cdot} \rightarrow$ stable products	k_6

The differential system can be derived as:

$$\frac{d[POOH]}{dt} = -\delta k_1[POOH]^{\delta} + k_3[POO^{\cdot}][PH] \tag{4.47}$$

$$\frac{d[P^{\cdot}]}{dt} = \alpha k_1[POOH]^{\delta} - k_2[P^{\cdot}][O_2] + k_3[POO^{\cdot}][PH] - 2k_4[P^{\cdot}]^2 - k_5[P^{\cdot}][POO^{\cdot}] \tag{4.48}$$

$$\frac{d[POO^{\cdot}]}{dt} = \beta k_1[POOH]^{\delta} + k_2[P^{\cdot}][O_2] - k_3[POO^{\cdot}][PH] - k_5[P^{\cdot}][POO^{\cdot}] - 2k_6[POO^{\cdot}]^2 \tag{4.49}$$

In the case of a thin sample (considered as a surface), the oxygen consumption rate (r_{ox}) is given by:

$$\frac{d[O_2]}{dt} = -k_2[P^{\cdot}][O_2] + k_6[POO^{\cdot}]^2 \tag{4.50}$$

In the case of deep layers of thick samples, it can be written:

$$\frac{\partial[O_2]}{\partial t} = D_{O_2}\frac{\partial^2[O_2]}{\partial z^2} - k_2[P^{\cdot}][O_2] + k_6[POO^{\cdot}]^2 \tag{4.51}$$

This is formally close to Equation 4.44. The numerical solution of the system allows one to obtain oxidation profiles in time and space that can be correlated with physico-chemical analysis [51].

4.4.3 Towards the Study of Coupling

Diffusivity, solubility, and permeability are commonly assumed to be constant for a given material, and this allows many simplifications for the numerical solution of the kinetic approach given above. However, since chemical ageing actually induces a strong modification of molar mass and subsequently of morphology in the case of semi-crystalline polymers [52], there is fact no reason to consider this simplifying assumption as valid. For example, in the case of PET photodegradation [53], it was observed that oxygen permeability decreases with the extent of oxidation. The decrease of OTR can be linearly correlated with the absorbance at 1785 cm^{-1}, ascribed to anhydrides as the main products of PET oxidation, and here attributable to PET crosslinking. Comparable results were obtained in irradiated polyethylene with a possible correlation between OTR and an increase in crystalline ratio [13], but contrasting results were recently observed by Marque et al. [54] for a EVOH/EVA/EVOH multilayer and Quelennec et al. [55]. The study of such coupling phenomena are worth further investigation in the future.

4.5 Conclusions

This chapter has described the basic principles of permeability of gases into polymers. Structural properties are reviewed, and allow a description of the effect of temperature on permeability for polymers in a rubbery or glassy state, as well as the effects of macromolecular architecture and crystalline ratio. Since those parameters can be modified by chemical ageing, these phenomena are described together with the possible coupling between ageing and oxygen diffusion.

References

[1] Masaro, L.; Zhu, X. X. Physical models of diffusion for polymer solutions, gels and solids. *Progress in Polymer Science* 1999, 24, 731–775

[2] Jung, J. K.; Kim, I. G.; Kim, K. T.; Ryu, K. S.; Chung K. S. Evaluation techniques of hydrogen permeation in sealing rubber materials. *Polymer Testing* 2021, 93, 107016

[3] Meares, P. The diffusion of gases through polyvinyl acetate. *Journal of the American Chemical Society* 1964, 76, 3415–3422

[4] Kanehashi, S.; Nagai, K. Analysis of dual-mode model parameters for gas sorption in glassy polymers. *Journal of Membrane Science* 2005, 253, 117–138

[5] Saberi, M.; Rouhi, P.; Teimoori, M. Estimation of dual mode sorption parameters for CO_2 in the glassy polymers using group contribution approach. *Journal of Membrane Science* 2020, 595, 117481

[6] Vieth, W. R.; Howell, J. M.; Hsieh, J. H. Review paper: Dual sorption theory. *Journal of Membrane Science* 1976, 1, 177–220

[7] Davis, E. G.; Rooney, M. L. Transport of sulphur dioxide inpolymers. *Kolloid-Zeitschrift und Zeitschrift für Polymere* 1971, 249, 1043–1050

[8] Vieth, W. R.; Frangoulis, C. S.; Rionda Jr. J. A. Kinetics of sorption of methane in glassy polystyrene. *Journal of Colloid and Interface Science* 1966, 22, 454–461

[9] Koros, W. J.; Paul, D. R.; Rocha, A. A. Carbon dioxide sorption and transport in polycarbonate. *Journal of Polymer Science: Polymer Physics Edition* 1976, 14, 687–702

[10] Haurat, M.; Tassaing, T. Dumon, M. FTIR in situ measurement of swelling and CO_2 sorption in acrylic polymers at high CO_2 pressures. *The Journal of Supercritical Fluids* 2022, 182, 105534

[11] Pino, M.; Duckett, R. A.; Ward, I. M. Single and mixed gas diffusion through polyethylene films. *Polymer* 2005, 46, 4882–4890

[12] Compañ, V.; Andrio, A.; López, M. L. Riande, E. Effect of annealing on the permeation characteristics of gases of coextruded LLDPE films. *Polymer* 1996, 37, 5831–5837

[13] Klepac, D.; Ščetar, M.; Baranović, G.; Galić, K.; Valić, S. Influence of high doses γ-irradiation on oxygen permeability of linear low-density polyethylene and cast polypropylene films. *Radiation Physics and Chemistry* 2014, 97, 304–312

[14] Marais, S.; Hirata, Y.; Langevin, D.; Chappey, C. O.; Nguyen, T. Q.; Metayer, M. Permeation and sorption of water and gases through EVA copolymers films. *Materials Research Innovations* 2002, 6, 79–88

[15] Hernandez, R. J. Effect of water vapor on the transport properties of oxygen through polyamide packaging materials. *Journal of Food Engineering* 1994, 22, 495–507

[16] Damian, C.; Espuche, E.; Escoubes, M. Influence of three ageing types (thermal oxidation, radiochemical and hydrolytic ageing) on the structure and gas transport properties of epoxy–amine networks. *Polymer Degradation and Stability* 2001, 72, 447–458

[17] Paul, D. R. Gas sorption and transport in glassy polymers. *Berichte der Bunsengesellschaft/Physical Chemistry Chemical Physics* 1979, 83, 294–302

[18] Maes, C.; te Molder, M.; Luyten, W.; Herremans, G.; Winckelmans, N.; Peeters, R.; Carleer, R.; Buntinx, M. Determination of the nitrogen gas transmission rate (N_2 GTR) of ethylene vinyl alcohol copolymer, using a newly developed permeation measurement system. *Polymer Testing* 2021, 93, 106979

[19] Weng, Y.; Ji, W.; Ye, C.; Dong, H.; Gao, Z.; Li, J.; Luo, C.; Ma, X. Simultaneously enhanced CO_2 permeability and CO_2/N_2 selectivity at sub-ambient temperature from two novel functionalized intrinsic microporous polymers. *Journal of Membrane Science* 2022, 644, 120086

[20] Yuan, Y.; Qiao, Z.; Xu, J.; Wang, J.; Zhao, S.; Cao, X.; Wang, Z.; Guiver, M. D. Mixed matrix membranes for CO_2 separations by incorporating microporous polymer framework fillers with amine-rich nanochannels. *Journal of Membrane Science* 2021, 620, 118923

[21] Cetina-Mancilla, E.; Ortencia González-Díaz, M.; Sulub-Sulub, R.; Zolotukhin, M. G.; González-Díaz, A.; Herrera-Kao, W.; Ruiz-Treviño, F. A.; Aguilar-Vega, M. Aging resistant, fluorinated aromatic polymers with ladderized, rigid kink-structured backbones for gas separations. *Journal of Membrane Science* 2022, 659, 120764

[22] Crank, J. Diffusion in a plane sheet, in *The Mathematics of Diffusion*. Clarendon Press, 1975, p. 44

[23] Haurat, M.; Tassaing, T. Dumon, M. FTIR in situ measurement of swelling and CO_2 sorption in acrylic polymers at high CO_2 pressures. *The Journal of Supercritical Fluids* 2022, 182, 105534

[24] Shelton, R. D.; Fletcher, W. H. The infrared spectrum and molecular constants of sulfur dioxide. *The Journal of Chemical Physics* 1953, 21, 2178–2183

[25] Khan, M. M.; Halder, K.; Shishatskiy, S.; Filiz, V. Synthesis and crosslinking of polyether-based main chain benzoxazine polymers and their gas separation performance. *Polymers* 2018, 10, 221

[26] Van Krevelen, D.W.; Te Nijenhuis, K. Properties determining mass transfer in polymeric systems (ch. 18), in *Properties of Polymers* (4th edition), Elsevier, 2009, pp. 655–702

[27] Park, J.; Ha, H.; Yoon, H. W.; Noh, J.; Park, H. B.; Paul, D. R.; Ellison, C. J.; Freeman, B. D. Gas sorption and diffusion in poly(dimethylsiloxane) (PDMS)/graphene oxide (GO) nanocomposite membranes. *Polymer* 2021, 212, 123185

[28] Messin, T.; Marais, S.; Follain, N.; Guinault, A.; Gaucher, V.; Delpouve, N.; Sollogoub, C. Biodegradable PLA/PBS multinanolayer membrane with enhanced barrier performances. *Journal of Membrane Science* 2020, 598, 117777

[29] Messin, T.; Follain, N.; Guinault, A.; Sollogoub, C.; Gaucher, V.; Delpouve, N.; Marais, S. Structure and barrier properties of multinanolayered biodegradable PLA/PBSA films: Confinement effect via forced assembly coextrusion. *ACS Applied Materials Interfaces* 2017, 9, 29101–29112

[30] Berens, A. R.; Hopfenberg, H. B. Diffusion of organic vapors at low concentrations in glassy PVC, polystyrene, and PMMA. *Journal of Membrane Science* 1982, 10, 283–303

[31] Fox, T. G.; Flory, P. J. Second-order transition temperatures and related properties of polystyrene. I. Influence of molecular weight. *Journal of Applied Physics* 1950, 21, 581–591

[32] Saberi, M.; Rouhi, P.; Teimoori, M. Estimation of dual mode sorption parameters for CO_2 in the glassy polymers using group contribution approach. *Journal of Membrane Science* 2020, 595, 117481

[33] Compañ, V.; Castillo, L. F.; Hernández, S. I.; Mar López-González, M.; Riande, E. Crystallinity effect on the gas transport in semicrystalline coextruded films based on linear low density polyethylene. *Journal of Polymer Science Part B: Polymer Physics* 2010, 48, 634–642

[34] Michaels, A. S.; Bixler, H. J. Flow of gases through polyethylene. *Journal of Polymer Science* 1961, 50, 413–439

[35] Compañ, V.; Ribes, A.; Diaz-Calleja, R.; Riande, E. Permeability of co-extruded linear low-density polyethylene films to oxygen and carbon dioxide as determined by electrochemical techniques. *Polymer* 1996, 37, 2243–2250

[36] Paulos, J. P.; Thomas, E. L. Effect of postdrawing on the permeability of gases in blown polyethylene film, *Journal of Applied Polymer Science* 1980, 25, 15–23

[37] Balakrishnan, P.; Gopi, S.; Sreerag, V. G.; Thomas, S. Mechanical and permeability properties of thermoplastic starch composites reinforced with cellulose nanofiber for packaging applications. *Journal of Siberian Federal University: Biology* 2019, 12, 287–301

[38] Nielsen, L. E. Models for the permeability of filled polymer systems. *Journal of Macromolecular Science Part A: Chemistry* 1967, 1, 929–942

[39] Balakrishnan, P.; Gopi, S.; Geethamma, V.G.; Thomas, S. Mechanical and permeability properties of thermoplastic starch composites reinforced with cellulose nanofiber for packaging applications. *Journal of Siberian Federal University: Biology* 2019, 12, 287–301

[40] Yang, F.; Manitiu, M.; Kriegel, R.; Kannan, R. M. Structure, permeability, and rheology of supercritical CO_2 dispersed polystyrene-clay nanocomposites. *Polymer* 2014, 55, 3915–3924

[41] Girard-Perier, N.; Dorey, S.; Gaston, F.; Girard, F.; Marque, S. R.; A. Dupuy, N. One-year ageing FTIR monitoring of PE/EVOH/PE film after gamma or electron beam irradiation. *Polymer Degradation and Stability* 2022, 195, 109790

[42] Hu, Y. S.; Prattipati, V.; Mehta, S.; Schiraldi, D. A.; Hiltner, A.; Baer, E. Improving gas barrier of PET by blending with aromatic polyamides. *Polymer* 2005, 46, 2685–2698

[43] *https://eval.kuraray.com/en-emea/products-services/about-eval-evoh/properties/barrier-to-oxygen/*

[44] Audouin, L.; Langlois, V.; Verdu, J.; de Bruijn, J. C. M. Role of oxygen diffusion in polymer ageing: Kinetic and mechanical aspects. *Journal of Materials Science* 1994, 29, 569–583

[45] Celina, M.; Wise, J.; Ottesen, D. K.; Gillen, K. T.; Clough, R. L. Correlation of chemical and mechanical property changes during oxidative degradation of neoprene. *Polymer Degradation and Stability* 2000, 68, 171–184

[46] Minervino, M.; Gigliotti, M.; Lafarie-Frenot, M. C.; Grandidier, J. C. The effect of thermo-oxidation on the mechanical behaviour of polymer epoxy materials. *Polymer Testing* 2013, 32, 1020–1028

[47] Richaud, E.; Colin, X.; Monchy-Leroy, C.; Audouin, L.; Verdu, J. Diffusion-controlled radiochemical oxidation of bisphenol A polysulfone. *Polymer International* 2011, 60, 371–381.

[48] Colin, X.; Marais, C.; Verdu, J. A new method for predicting the thermal oxidation of thermoset matrices: Application to an amine crosslinked epoxy. *Polymer Testing* 2001, 20, 795–803

[49] Grandidier, J. C.; Olivier, L.; Lafarie-Frenot, M. C.; Gigliotti, M. Modeling the pressure dependent solubility in a thermoset resin for simulating pressure accelerated thermo-oxidation tests. *Mechanics of Materials* 2015, 84, 44–54

[50] Celina, M.; Wise, J.; Ottesen, D. K.; Gillen, K. T.; Clough, R. L. Correlation of chemical and mechanical property changes during oxidative degradation of neoprene. *Polymer Degradation and Stability* 2000, 68, 171–184

[51] Ernault, E.; Dirrenberger, J.; Richaud, E.; Fayolle, B. Prediction of stress induced by heterogeneous oxidation: Case of epoxy/amine networks. *Polymer Degradation and Stability* 2019, 162, 112–121

[52] Fayolle, B.; Richaud, E.; Colin, X.; Verdu, J. Review: Degradation-induced embrittlement in semi-crystalline polymers having their amorphous phase in rubbery state. *Journal of Materials Science* 2008, 43, 6999–7012

[53] Gardette, J. L.; Colin, A.; Trivis, S.; German, S.; Therias, S. Impact of photooxidative degradation on the oxygen permeability of poly(ethyleneterephthalate). *Polymer Degradation and Stability* 2014, 103, 35–41

[54] Gaston, F.; Dupuy, N.; Girard-Perier, N.; Marque, S. R. A.; Dorey, S. Investigations at the product, macromolecular, and molecular level of the physical and chemical properties of a irradiated multilayer EVA/EVOH/EVA film: Comprehensive analysis and mechanistic insights. *Polymers* 2021, 13, 2671

[55] Quelennec, B.; Duan, Z.; Delannoy, R.; Gay, N.; Briffaut, M.; Tognetti, V.; Delpouve, N.; Delbreilh, L.; Bredif, L.; Duthoit, A.; Richaud, E. Effect of physical and chemical ageing on barrier properties of epoxy coating. *Construction and Building Materials* 2023, 409, 133908

5 Water Sorption Behavior in Polymers: Kinetic and Thermodynamic Approaches

Stéphane Marais and Kateryna Fatyeyeva

5.1 Introduction

Due to their large range of physico-chemical properties, polymers are the materials of choice when designing systems combining lightness, high performance, and low cost. The development of technological manufactory process (i. e. 3D printing, plastics engineering process, etc.) allows polymer modularity in shape, and structures with adaptive and functional properties at multiple scales. Also, blending or alloying polymers, multilayering, grafting interactive chemical groups, surface treatment, and introducing nanofillers or synthetic/natural reinforcements make it possible to achieve significant improvements in the material's performance in terms of barrier, selective, mechanical, thermal, conductive, and optical properties.

On the other hand, as polymers can be sensitive to severe environmental conditions, it is important to emphasize that during sorption/diffusion (owing to possible chemical modifications and/or relaxation phenomena) the material's properties can be altered and premature aging can take place. In particular, such phenomena can occur during the absorption of water, causing hydrolysis reactions and structural/morphological changes. Thus, deep analysis of the sorption/migration of a penetrant in polymers allows us to study the behavior of the polymer materials and the possible evolution of a material over time during an absorption situation.

Sorption and permeation processes provide complementary information about transport properties resulting from the different boundary conditions. Indeed, they aim to study the material's ability to dissolve gases and/or vapors (i. e. sorption) or to let them to pass (in the case of permeation). The advantage of coupling these two approaches lies in the determination of the transport parameters (i. e. solubility, diffusivity, and permeability) at different time scales (namely, first half-sorption when permeation is established, and second half-sorption when the system reaches a steady state and sorption equilibrium). This reveals structural heterogeneities (owing to surface treatments, cross-linking irregularities through the thickness, inclusion of nanofillers, etc.) and the dependency of the diffusion coefficient over time (aggregation, relaxation, chemical reactions, etc.). They also offer the possibility of probing matter by changing the nature, size, and activity of the diffusing molecules (as single species, or in a mixture), and the temperature.

This chapter aims to provide a general overview of the kinetic and thermodynamic aspects of the water sorption in polymers from a fundamental and experimental point of view.

5.2 Theoretical Background

Small molecules may be dissolved and may diffuse in dense polymers (Figure 5.1a). This phenomenon is generally characterized by sorption and permeation in an isothermal medium and under the following conditions:

- The polymer material is in the film form and its thickness L is negligible compared with the other two dimensions (i.e. weight and length). A is the section of the film exposed to the diffusion
- The polymer is initially "penetrant-empty" (or in the case of desorption, "penetrant-filled")
- The penetrant diffuses from a source phase, in which its activity a (in the case of liquid or vapor) or its pressure p (in vase of gas or vapor) is kept constant.

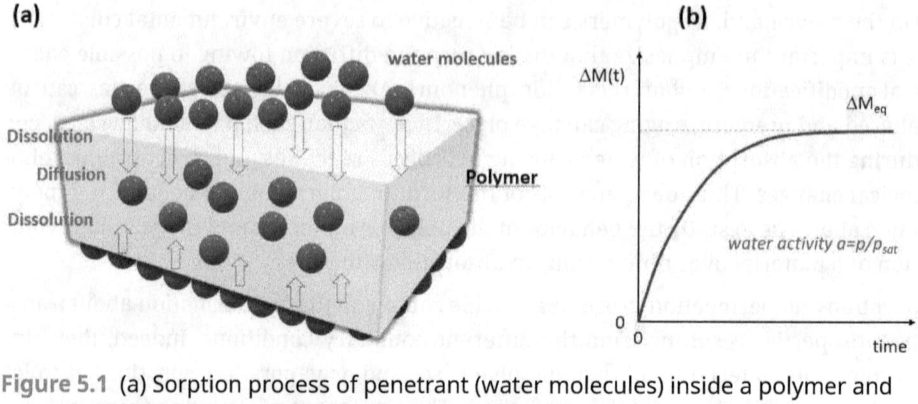

Figure 5.1 (a) Sorption process of penetrant (water molecules) inside a polymer and (b) mass variation of sorbed molecules as a function of time

During absorption process, the film is immersed in the source phase. The quantity of the penetrant $\Delta M(t)$ accumulated in the film is determined as a function of time. At the end of the kinetic phase, $\Delta M(t)$ approaches the equilibrium value ΔM_{eq}. Taking into account the material unit volume, ΔM_{eq} becomes the equilibrium concentration C_{eq}, variation of which with a or p represents the exchange isotherm. An absorption process can be followed by desorption. In this case, the "filled" film can be placed in a vacuum chamber or subjected to a scavenging gas, and the kinetics of penetrant release can be measured.

5.2.1 Sorption Kinetics

Sorption kinetics can be analyzed by immersing a dry film in a medium, in which the water activity a is kept constant. This medium can be liquid water ($a = 1$, by reference to the pure body under 1 atm), a saline solution (generally $a < 1$) or a humid atmosphere ($a = p/p_0$, i.e. relative humidity or vapor tension p related to the saturating vapor pressure p_0). The mass variation of water inside the polymer as a function of time $\Delta M(t)$ is measured until it reaches the equilibrium state corresponding to ΔM_{eq} (Figure 5.1b).

At equilibrium, the mass fraction m_{eq} and the water concentration C_{eq} in the polymer can be expressed by the activity a and a solubility coefficient σ:

$$C_{eq} = \frac{m_{eq}\rho}{0.018} = \sigma a \tag{5.1}$$

Where:

ρ = density of the wet film in g cm^{-3}

0.018 = mass of 1 mmol of water in g

If the polymer absorbs ΔM_{eq} g of water per M_0 g of polymer, the gain in mass at equilibrium (in %) is 100 $\Delta M_{eq}/M_0$ and the mass fraction is $m_{eq} = \Delta M_{eq}/(M_0 + \Delta M_{eq})$. Here C_{eq} and σ are expressed in mmol of water/cm^3 of wet film. An analysis of the equilibria as a function of the water activity makes it possible to establish the sorption isotherm as m_{eq} or $C_{eq} = f(a)$. For each activity, the sorption kinetics is characterized by the transient regime before reaching the equilibrium. The sorption kinetics are usually expressed as:

$$\frac{\Delta M_t}{\Delta M_{eq}} = kt^n \tag{5.2}$$

5.2.1.1 Fickian Sorption of Type I (n = ½)

In this case, the penetrant diffusion rate is lower than the rate of molecular chain segments, the sorption rate is not dependent on the polymer swelling rate, and the sorption equilibrium is rapidly reached [1]. As the polymer relaxation is much faster than the solvent diffusion, this Fickian behavior is often observed in polymers with high flexibility, i. e. in polymers for which the measuring temperature is above the polymer glass transition temperature T_g.

During diffusion, mass transfer is governed by the two Fick's laws, which allow linking the local flux density $J(x, t)$ with the local concentration gradient $C(x, t)$ [2]:

$$J(x,t) = -D\frac{\partial C(x,t)}{\partial x} \tag{5.3}$$

and the evolution of the local concentration gradient to the flux density:

$$\frac{\partial C(x,t)}{\partial t} = -\frac{\partial J(x,t)}{\partial x} \tag{5.4}$$

These relationships include a diffusion coefficient D. Usually D is expressed in $cm^2\,s^{-1}$, x is in cm, C is in $mol\,L^{-1}$ (or $mmol\,cm^{-3}$) and J is in $mmol\,cm^{-2}\,s^{-1}$. When D is constant, Equation 5.4 becomes:

$$\frac{\partial C(x,t)}{\partial t} = D\frac{\partial^2 C(x,t)}{\partial x^2} \tag{5.5}$$

Fick's laws assume that the penetrant molecules freely diffuse through the dense film without chain rearrangements. The following hypotheses should be taken into consideration:

- The polymer material is homogeneous
- The film thickness is negligible compared to the other film dimensions
- The effect of swelling is negligible
- The diffusion is not time-dependent
- The sorption equilibrium at the film surface is almost instantaneous.

The polymer film is usually considered to be inert towards permeant species. A dense film corresponds to a medium composed of penetrant-scale gaps, the size and position of which are changing randomly and continuously in time. It therefore shows typical characteristics of stochastic geometry, and may be characterized by the Brownian motion and the percolation theory.

The sorption and permeation kinetics that obey Fick's laws are distinguished by the boundary conditions. The equations of the following paragraphs can be applied to the polymer film assuming instantaneous establishment of the interfacial equilibrium.

5.2.1.1.1 Principle

A dry film of constant thickness L is placed in a medium with a constant water activity and an instantaneous interfacial equilibrium. The boundary conditions imposed upon the system are following:

$$\text{at } t = 0 \tag{5.6}$$

$$C(x,0) = 0 \quad \forall x \in]0, L[$$

$$\text{at } x = 0 \text{ or } L$$

$$C(0,t) = C(L,t) = C_{eq}, \forall t$$

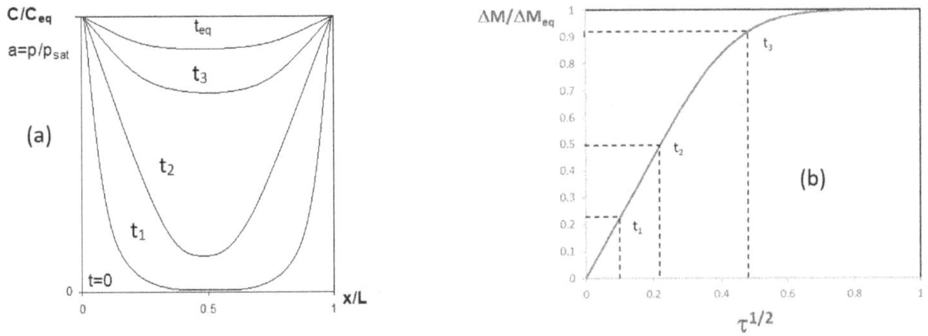

Figure 5.2 (a) Sorption concentration profile and (b) sorption kinetics of reduced mass as a function of dimensionless time $\tau^{1/2}$

The evolution of the concentration profiles (Figure 5.2a) and the absorbed water quantity can be determined from Equation 5.3 and Equation 5.4 [2, 3]. The absorbed water quantity can be expressed by a number n (in mmol) or by a mass ΔM (in g):

$$n = \frac{\Delta M}{0.018} = A \int_{x=0}^{L} C(x,t)\, dx \tag{5.7}$$

Where:

A = area of each side of the film (in cm^2)

At equilibrium, n and ΔM become equal to n_{eq} and ΔM_{eq}, so that:

$$n_{eq} = \frac{\Delta M_{eq}}{0.018} = A C_{eq} L \tag{5.8}$$

D is the penetrant local diffusion coefficient, for which Frisch [4] distinguishes two Fickian mechanisms: when D is constant (type A) and when D is a function of the concentration (type B).

5.2.1.1.2 Constant Diffusion Coefficient

In the case of a constant diffusion coefficient (i. e. Fickian mechanism of type A), the time can be expressed in the form of a dimensionless variable:

$$\tau = \frac{Dt}{L^2} \tag{5.9}$$

Therefore, the concentration profiles and the quantity of absorbed water can be obtained by analytical solutions in the form of a series.

Concentration profile

$$\frac{C}{C_{eq}} = \text{erfc}\frac{x/L}{2(\tau)^{1/2}} + \text{erfc}\frac{1-x/L}{2(\tau)^{1/2}} + \sum_{n=0}^{\infty}(-1)^n\left(\text{erfc}\frac{n+x/L}{2(\tau)^{1/2}} + \text{erfc}\frac{(n+1)-x/L}{2(\tau)^{1/2}}\right) \tag{5.10}$$

$$\frac{C}{C_{eq}} = 1 - \frac{4}{\pi}\sum_{n=0}^{\infty}\frac{(-1)^n}{2n+1}\cos\left[(2n+1)\pi\frac{x}{L}\right]e^{(-(2n+1)^2\pi^2\tau)} \tag{5.11}$$

Equation 5.10 converges more rapidly during the first half-sorption until time $\tau_{1/2} = 0.0492$, for which $\Delta M/\Delta M_{eq} = 0.5$. Beyond this value, Equation 5.11 is preferably used.

Amount of sorbed water

For short times ($\tau \leq \tau_{1/2}$):

$$\frac{n}{n_{eq}} = 4\left(\frac{\tau}{\pi}\right)^{1/2}\left(1 + 2(\pi)^{1/2}\sum_{n=1}^{\infty}(-1)^n\text{ierfc}\frac{n}{2(\tau)^{1/2}}\right) \tag{5.12}$$

and for longer times ($\tau \geq \tau_{1/2}$):

$$\frac{n}{n_{eq}} = \frac{\Delta M}{\Delta M_{eq}} = 1 - \frac{8}{\pi^2}\sum_{n=0}^{\infty}\frac{e^{[-(2n+1)^2\pi^2\tau]}}{(2n+1)^2} \tag{5.13}$$

The concentration profiles (Figure 5.2a) and the absorbed quantity (Figure 5.2b) are calculated from these different expressions.

During the first half-sorption ($\tau \leq \tau_{1/2}$) (Figure 5.3a), Equation 5.10 and Equation 5.12 converge very quickly. In particular, the water quantity retained by the film changes practically according to:

$$\frac{\Delta M}{\Delta M_{eq}} = \left[\frac{4}{L}\sqrt{\frac{D}{\pi}}\right]\sqrt{t} \tag{5.14}$$

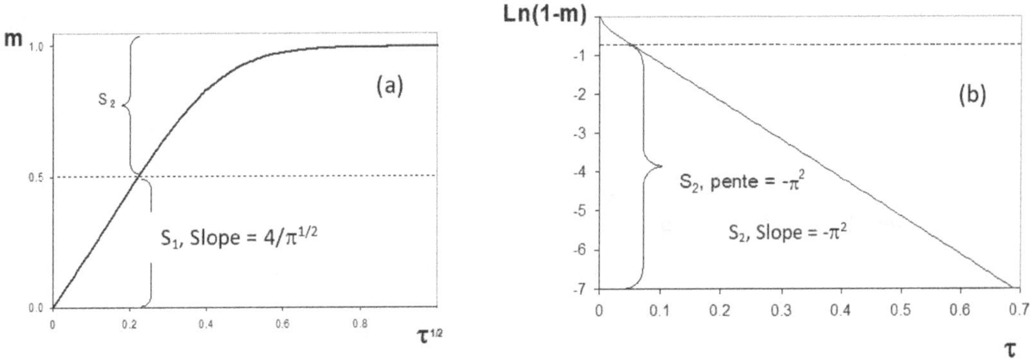

Figure 5.3 Sorption kinetic curves as a function of (a) $\tau^{1/2}$ for the first half-sorption and (b) τ for the second half sorption

For the second half-sorption ($\tau \geq \tau_{1/2}$) (Figure 5.3b), Equation 5.11 and Equation 5.13 are the most suitable, and the evolution of the retained water quantity may be expressed by:

$$\ln\left(1 - \frac{\Delta M}{\Delta M_{eq}}\right) = -\ln\frac{\pi^2}{8} - \frac{\pi^2 Dt}{L^2} \tag{5.15}$$

These two simple laws allow immediate calculation of the diffusion coefficient D, and they can be applied for absorption ($m_a = \Delta M/\Delta M_{eq}$) as well as for desorption ($m_d = 1 - \Delta M/\Delta M_{eq}$). In practice, the equilibrium is reached beyond a time $t_{0.99}$, for which the sample absorbs 99% of water.

5.2.1.1.3 Concentration-Dependent Diffusion Coefficient

The previous theoretical curves (Figure 5.2 and Figure 5.3) were established for the ideal case, when the diffusion coefficient is constant. The experimental reality can be different, and there are many examples in the literature when the penetrant diffusion coefficient is a function of its concentration in the polymer. However, at the beginning of the sorption, the kinetics always follows a law in $t^{1/2}$ and, thus, an average diffusion coefficient $\langle D \rangle$ can be deduced. The overall curve shape follows the dependence $D = f(C)$.

For the first half-sorption, Equation 5.14 can be generalized to the case of the concentration-dependent diffusion coefficient, so that the diffusion coefficient D_1 is determined as:

$$D_1 = \left(\frac{\sqrt{\pi}L}{4}\right)^2 \left(\frac{dm(t)}{d\sqrt{t}}\right)^2 \tag{5.16}$$

For the second half-sorption, it is common to determine the diffusion coefficient D_2 from Equation 5.15:

$$D_2 = -\left(\frac{L}{\pi}\right)^2 \frac{\mathrm{d}(\ln(1 - m(t))}{\mathrm{d}t} \tag{5.17}$$

Specific case $D = D_0 \exp(\gamma C)$

In the literature, the concentration-dependent diffusion coefficient is usually expressed by the exponential law:

$$D = D_0 \exp(\gamma C) \tag{5.18}$$

Where:

D_0 = limiting diffusion coefficient at zero concentration

C = permeant concentration in the polymer

It is generally applied to penetrants with a strong plasticizing effect, such as organic compounds vapor in hydrophobic polymers or water vapor in hydrophilic polymers (Fickian type B). In a rubbery polymer ($\gamma > 0$ [2, 5]), the γ parameter often being considered as a plasticization coefficient. This law, which refers to the "free volume" models, will be discussed in detail in Section 5.3.2.1.

Considering this law and the boundary conditions of Equation 5.6, the diffusion coefficient can vary from D_0 ($C = 0$) to $D_M = D_0 \exp(\gamma C_{eq})$. By applying Runge–Kutta method [6], Fick's law can be solved. Thus, taking the dimensionless time $\tau_0 = D_0 t/L^2$, a series of curves $m = f(\tau_0^{1/2})$ (Figure 5.4a) obtained for various γC_{eq} values is characterized by slope β_0, which is constant during the first half-sorption. In agreement with Equation 5.14 and Equation 5.16 and by using the time-scale change [7]:

$$\tau_1 = \left(\frac{\sqrt{\pi}\beta_0}{4}\right)^2 \tau_0 = \frac{D_1 t}{L^2} = \tau_0 \frac{D_1}{D_0} \tag{5.19}$$

A new series of curves is obtained, which merged together in the linear part of the first half-sorption, so the effect of γC_{eq} appears only in the second half-sorption (Figure 5.4b).

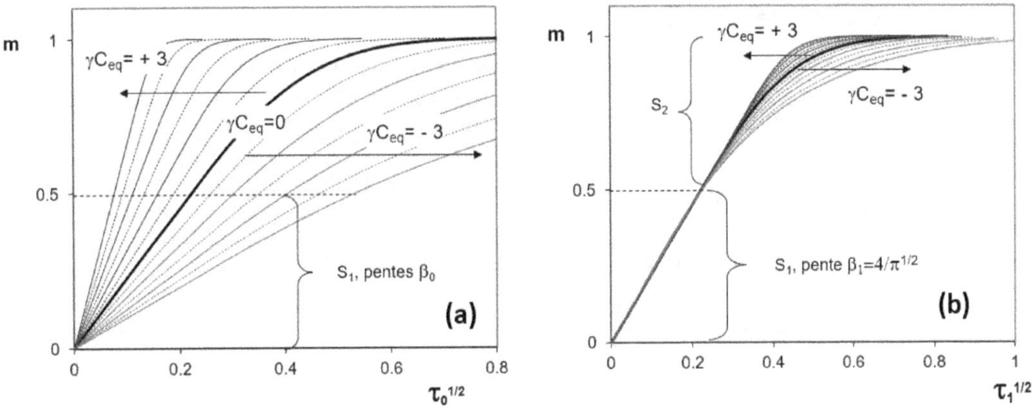

Figure 5.4 Sorption kinetic curves assuming $D = D_0 \exp(\gamma C_{eq})$ and plotted for various γC_{eq} values as a function of (a) $\tau_0^{1/2}$ and (b) $\tau_1^{1/2}$

From the linear regression of Figure 5.4b, the slope β as a function of γC_{eq} can be deduced:

$$\beta = \frac{\mathrm{d}\ln(1-m)}{\mathrm{d}\tau_1} = -\pi^2 \frac{D_2}{D_1} \tag{5.20}$$

and a relationship between γC_{eq} and the D_2/D_1 ratio can be established.

Absorption

Regarding absorption (i. e. $D_1 = D_1^a$, $D_2 = D_2^a$ and $\beta = \beta^a$) in the range of $-3 \leq \gamma C_{eq} \leq +3$, the following expressions could be used for:

- The first half-sorption:

$$\ln \frac{D_1^a}{D_0} \approx 0.026(\gamma C_{eq})^2 + 0.632\gamma C_{eq} \tag{5.21}$$

- The second half-sorption:

$$\gamma C_{eq} \approx -2.1752 - 0.2754\beta_a - 0.0001\beta_a^3 + \frac{6.46236}{\beta_a} \tag{5.22}$$

with

$$\beta_a = -\pi^2 \frac{D_2^a}{D_1^a}$$

Absorption–desorption

The concentration dependence of D is also underlined by the difference in absorption and desorption kinetics. For the first half-sorption, $D_1 = D_1^{d}$, so Equation 5.21 can be written as:

$$\ln \frac{D_1^{d}}{D_0} \approx 0.026(\gamma C_{eq})^2 + 0.368\, \gamma C_{eq} \tag{5.23}$$

The γC_{eq} value can then be deduced as follows:

$$\gamma C_{eq} \approx \frac{1}{0.264} \ln \frac{D_1^{a}}{D_1^{d}} \tag{5.24}$$

5.2.1.2 Fickian Sorption of Type II (n = 1)

The diffusion rate of penetrant molecules is much higher than the mobility rate of macromolecular chains. In the case of organic vapor sorption in glassy polymers, a strong plasticization effect can be induced, so that the sorption kinetics is highly dependent on the polymer swelling rate. Thus, the permeation process is considered as a sorption-controlled process, in which the diffusion rate is rapid and the sorption is the key step.

When the sorption and desorption are time-dependent (i. e. when during the sorption the relaxation occurs as a result of the polymer swelling), Fick's laws are not appropriate for the mass transfer modelization. Therefore, Kholer et al. [8] developed the "parallel exponential kinetics" (PEK) model to model the water sorption kinetics of cellulosic fibers. This model considers the water vapor sorption as a first-order formation of sorbate S_1 and S_2 by two independent parallel processes characterized by two-time constants k_1 and k_2. So, two different sorption sites with a limited sorption capacity storage of an equilibrium mass M_∞ should be taken into account. The combination of both sorption processes is expressed as the sum of a slow and a rapid kinetic process, in which the characteristic time t (to reach 63% of M_∞) is used instead of the time constant, according to:

$$M = M_{\infty,S_1}\left(1 - \exp\left(-\frac{t}{\tau_{S_1}}\right)\right) + M_{\infty,S_2}\left(1 - \exp\left(-\frac{t}{\tau_{S_2}}\right)\right) \tag{5.25}$$

Where:

M = mass uptake

τ_{S_1} and τ_{S_2} = characteristic times of the rapid and slow kinetics (respectively)

M_{∞,S_1} and M_{∞,S_2} = mass uptakes at infinite time corresponding to the saturation for each sorption process

The rapid and slow sorption sites can be related to different amorphous zones, external/internal fiber surfaces, and direct/indirect sorption [9]. The direct sorption of water molecules onto external surfaces and amorphous regions can be assumed to be rapid, whereas the indirect sorption onto the inner surfaces and crystallites can be related to the slow process.

5.2.1.3 Non-Fickian Sorption of Type III (½ < *n* <1)

This type of sorption, known as anomalous diffusion, occurs when the molecule's diffusivity is comparable to the polymer relaxation. It means that the diffusion rate is comparable to the sorption rate. However, any theory can explain correctly this behavior.

5.2.2 Sorption Isotherms

The sorption isotherm deduced from the equilibrium state of the different sorption kinetics at various relative humidity (RH) or water activity ($a = p/p_{sat}$), or vapor pressure p is represented by the mass gain $\Delta M/\Delta M_{eq}$ (%) or equilibrium water concentration C_{eq} (g/g) as a function of RH, a, or p (Figure 5.5).

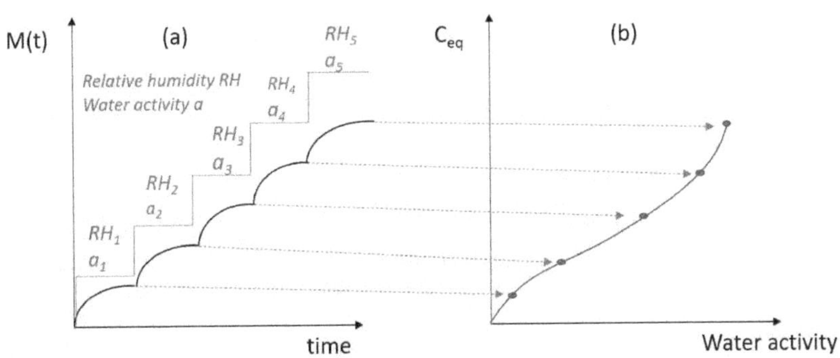

Figure 5.5 (a) Sorption kinetics curves at different relative humidity (water activity) and (b) sorption isotherm

Various shapes of sorption isotherm curves can be found in the literature. Using numerous models (empirical and semi-empirical) to describe the sorption isotherm behavior, Rogers (types I, II, III, IV) [10] and Brunauer–Emmett–Teller (BET) classifications are usually used to model the gas and vapor sorption in polymer materials. In most cases, five sorption isotherm curves can be obtained, as shown in Figure 5.6.

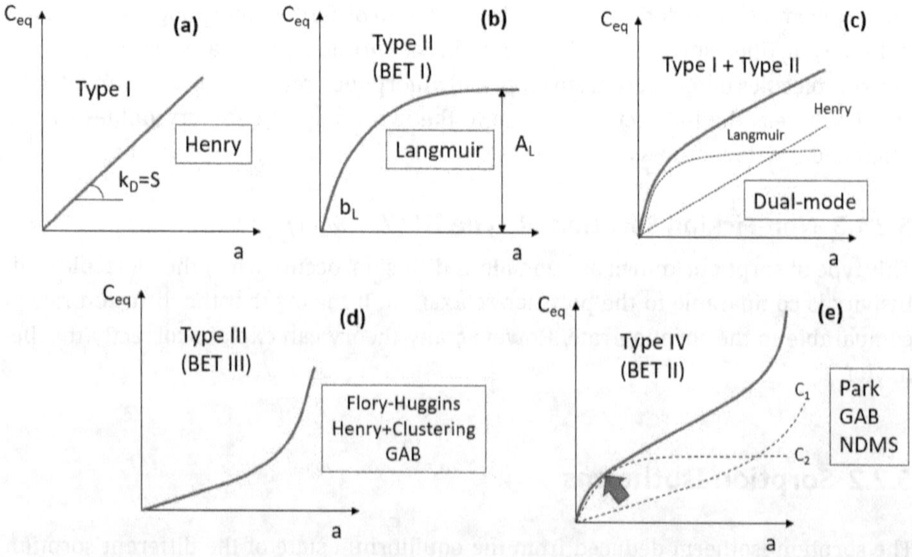

Figure 5.6 Main types of sorption isotherm curves in polymers

5.2.2.1 Sorption Modes

Henry sorption – Rogers type I

This sorption mode is considered as the ideal case, when the penetrant–penetrant interactions and the penetrant–polymer interactions are weak. In general, this sorption mode takes place in the case of a permanent gas–rubber polymer system [11]. The absorption of the penetrant molecules occurs in a uniform and non-specific manner in the polymer matrix, and the solubility coefficient is quite low.

In this case, the solubility coefficient denoted k_D (Henry's constant) (Figure 5.6a) takes into account the ability of the penetrant "to be condensed" in the polymer matrix. The concentration is given as follows:

$$C_{eq} = k_D a \tag{5.26}$$

Langmuir sorption – Rogers type II/BET type I

The Langmuir sorption originates from the gas adsorption into microporous solids. The physisorption takes place on specific sites, and so the sorbed molecules are immobilized. According to Langmuir, there is competition between the condensation and the evaporation on the solid surface [12]. At equilibrium, when the Langmuir curve reaches a plateau (Figure 5.6b) (meaning all the sites are saturated), the condensation flux (i. e. the association rate v_a) and the evaporation flux (i. e. the dissociation rate v_d) are equal, so that the fraction of occupied surface θ can be deduced:

$$\theta = \frac{b_L p}{1 + b_L p} \tag{5.27}$$

Where:

b_L = Langmuir constant (positive), which corresponds to the ratio of the rate constants ($b_L = K_a/K_d$)

p = pressure

The equilibrium concentration is given as:

$$C_{eq} = \frac{C_L b_L a}{1 + b_L a} \tag{5.28}$$

Where:

C_L = mean concentration of Langmuir sites

b_L = affinity constant of penetrant molecules for these sites

This type II isotherm is given for the monolayer adsorption in microporous solids with relatively small external surfaces.

The average Langmuir concentration C_L is related to the ion exchange capacity. In the case of water and ion exchangers (a special case), the more ions there are on the exchange sites, the higher the C_L value. The Langmuir adsorption occurs, for example, in the case of the sorption of the dyes by ionic polymers or polymers with porous particles (carbon black, gel silica, etc.) [13].

The Jovanovic theory of adsorption [14] is quite similar to the Langmuir theory. However, in this case the surface available for adsorption is not proportional to $(1 - \theta)$, but slightly higher, as the evaporation of adsorbed molecules is taken into account. Thus, the number of moles adsorbed by the solid is expressed as:

$$n = n_0(1 - e^{-Kp}) \tag{5.29}$$

Dual-mode sorption – Types I + II

This type of isotherm is observed in the case of the sorption of permanent gases (such as carbon dioxide) in glassy polymers (Figure 5.6c). Many authors [11, 15] describe the "dual-mode" model supposing that the gas dissolved in the glassy polymer consists of two distinct molecular populations – namely, molecules dissolved in a limited number of microvoids that are fixed or pre-existing in the polymer matrix (i. e. the Langmuir sorption), and molecules dissolved in the polymer by a non-specific dissolution process (i. e. the Henry sorption):

$$C_{eq} = C_{Langmuir} + C_{Henry} = \frac{C_L b_L a}{1 + b_L a} + k_D a \tag{5.30}$$

Flory–Huggins sorption – Type III (BET III)

The sorption of organic vapors in elastomeric or rubbery materials (for example, chloroform sorption in silicone [16]), but also the sorption of plasticizers (for example, the sorption of vinyl chloride monomer in PVC [17]) is often described by the Flory–Huggins (FH) theory (Figure 5.6d). This theory is based on the lattice model of mixing [12, 18, 19], and is used to explain the organization of the polymer chains and solvent molecules, taking into consideration three types of interactions: (1) polymer segment–polymer segment; (2) solvent molecule–solvent molecule; and (3) polymer segment–solvent molecule. The polymer–solvent solution is arbitrarily divided into a set of N_0 cells: each cell can be occupied either by one polymer segment or by one solvent molecule. One polymer chain is defined as containing N segments with the same molar volume as the solvent volume.

$$\ln a = \ln \phi + (1 - \phi) + \chi(1 - \phi)^2 \tag{5.31}$$

Where:

ϕ = volume fraction occupied by solvent

$(1 - \phi)$ = volume fraction occupied by the polymer

χ = Flory–Huggins parameter of mutual interaction defining the polymer miscibility with solvent

Equation 5.31 is deduced from the free energy mixing assuming that polymer chains are sufficiently long (i. e. high N value). In a generalized Flory–Huggins theory, the χ parameter, initially assumed to be constant, is expressed as the product of concentration-dependent and temperature-dependent terms [20, 21]. A good plasticizer or solvent has a low χ value. The interaction force between the plasticizer molecules and polymer macromolecules should be greater than the interaction force between the plasticizer molecules themselves. When the plasticizing phenomenon occurs, a concentration-dependent diffusion coefficient is usually observed due to the increase of free volume, which leads to an increase of D with concentration C according to the well-known exponential law (Equation 5.18).

In the case of water sorption isotherm with a convex shape, three other models based on the clustering formation are preferentially used, namely the two-mode sorption Henry–clustering model, the engaged species induced clustering (ENSIC) model, and the Guggenheim–Anderson–de Boer (GAB) model. In contrast to the Flory–Huggins theory, as solvent–solvent interactions become predominant, the diffusion coefficient decreases with the water concentration rising owing to the steric hindrance induced by the size of the cluster, thus reducing the water mobility. Perrin et al. showed that the water sorption isotherm in hydrophilic cellulose triacetate could also be well described by the Flory–Huggins theory, but only at activities less than 0.7 [22]. According to this theory, when a penetrant is added to the material solute, the enthalpy changes as the penetrant–penetrant and solute–solute interactions are replaced by the penetrant–solute interactions. It should be mentioned that the interactions between the penetrant and solute are assumed to be constant whatever the penetrant activity is. However, the clustering phenomenon is not clearly explained, as deviation from the Flory–Huggins thermodynamic theory is observed at high penetrant activity (≥0.7).

Two-mode sorption: Henry–clustering (HC)

The two-mode sorption Henry–clustering model exhibits a curve shape similar to that of Flory–Huggins (Figure 5.6d) that is a quasi-linear curve at low and intermediate activity followed by a convex curvature at higher activity ($a > 0.5$). This two-mode sorption model is used in the case of water sorption in hydrophobic materials [23]. In the case of a binary system, the penetrant–penetrant and penetrant–material interactions are found to be dependent on the penetrant activity value. Thus, it is assumed that two species of absorbed water molecules contribute to the water concentration in the polymer: one of these species follows Henry's law at low and intermediate water activity ($0 < a \leq 0.7$) (i. e. molecules are randomly sorbed with no specific interaction with a polymer), and the second species follows the water clustering at higher activity ($a > 0.7$) (i. e. some of the sorbed molecules from the Henry's law sorption forms a cluster due to the predominant penetrant–penetrant interactions at high water concentration). Therefore, taking into account these combined sorption modes, the water sorption isotherm may be expressed as follows:

$$C_{\text{eq}} = C_{\text{Henry}} + C_{\text{cluster}} = k_{\text{D}} a + n K_{\text{c}} k_{\text{D}}^n a^n \tag{5.32}$$

Where:

C_{eq} ($g_{\text{water}}/g_{\text{polymer}}$) = total water concentration

C_{Henry} and C_{cluster} = water concentration from the Henry's law sorption and from the clustered species (respectively)

a is water activity, k_{D} = Henry's solubility coefficient representing the affinity of water molecules to the sample

n = mean number of water molecules per cluster

K_{a} = equilibrium constant of the clustering reaction $nH_2O \leftrightharpoons (H_2O)_n$

ENSIC model

The mechanistic ENSIC model takes into account the cluster formation, and considers the sorption of vapor molecules on either a polymer segment or on a previously sorbed molecule [1, 24].

$$\phi = \frac{[e^{(k_s - k_p)a} - 1]}{(k_s - k_p)/k_p} \tag{5.33}$$

Where:

ϕ = solvent volume fraction

k_s = affinity parameter between solvent and previously sorbed molecule

k_p = affinity parameter between solvent and the polymer segment

This model is mainly used in the case of the sorption of polar solvents in non-polar and polar systems.

BET III model

Among the numerous models proposed to describe the isotherm curve of convex shape, the BET model of type III can be considered. According to this model, gas molecules are physically adsorbed on a solid surface in form of multilayers. In addition, there are no interactions between each adsorption layer, and the Langmuir theory can be applied to each layer. In the case of type III model, the isotherm is convex to the pressure (activity) axis, a multilayer adsorption process occurs owing to strong lateral interactions between adsorbed molecules in comparison with interactions between adsorbent surface and adsorbate. For example, Vopicka et al. [20] revealed that the GAB theory of multilayer adsorption is more appropriate to describe the sorption of vapors (water, methanol, acetone, methyl acetate, dimethyl carbonate, and *tert*-butyl methyl ether) in cross-linked PDMS than the Flory–Huggins theory with a concentration-dependent interaction parameter and the ENSIC model.

GAB model

The Guggenheim–Anderson–de Boer (GAB) model, derived from the BET theory, is often used to fit water sorption isotherms (concave and convex curves), especially in the field of food products (Figure 5.6d) [25, 26]. In fact, the GAB model considers a limited number of layers of sorbed molecules with no lateral interactions. The molecules in the so-called multilayer have interactions with the penetrant, which range in energy levels between the interactions of the monolayer molecules and the bulk liquid. Consecutive water molecule layers may possess bulk liquid properties. However, the GAB model differs from the BET II model by the hypothesis of the evaporation–condensation properties from the second layer and, thus, by introducing a new parameter K (associated with the water molecules sorbed on the multilayers) and by replacing the constant C_{BET} by the Guggenheim constant C_G:

$$C_{eq} = C_m \frac{C_G K a}{(1 - Ka)(1 - Ka + C_G K a)} \tag{5.34}$$

The GAB model corresponds to the BET model, assuming that the adsorption energy of the upper layers are lower than that of the first layer, so that $K < 1$. The water molecules beyond to the monolayer are not structured in a multilayer, but have the same characteristics as the molecules in a liquid bulk.

Sigmoidal sorption isotherm

The water sorption in hydrophilic polymers (for example, polyvinylalcohol (PVA), polyacrylic acid (PAA)) and derivative of cellulosic materials leads to isotherm curve of sigmoidal shape (Figure 5.6e). The first theoretical equation used for this kind of isotherm is the well-known BET relationship. Such isotherms are classified as type IV according to the Rogers classification.

The BET II isotherm, which represents an unrestricted monolayer–multilayer adsorption, is a combination of the Langmuir (first part of the curve, i. e. concave to the activity axis) and Flory–Huggins (second part, i. e. convex to the activity axis) isotherms. When the monolayer formation is completed, the multilayer adsorption begins. This moment is located at the beginning of the almost linear middle section of the isotherm curve (labelled by the red arrow in Figure 5.6e).

New-dual mode sorption (NDMS)

For sigmoidal isotherms, convex isotherms (Langmuir or dual-mode type), or convex isotherms (Flory–Huggins type) of vapor sorption in glassy polymers, Feng [27] developed the NDMS model (Figure 5.6e). This model is based on four assumptions: (1) the sorption site in a glassy polymer can be divided into two different species – one in the polymer matrix and the other one in the microvoids; (2) all sorbed molecules in the matrix have the same partition function; (3) there is the GAB sorption in the microvoid region of the polymer; (4) the molecules in the layers over the first layer in the microvoids have the same partition function as molecules in the matrix region. The equilibrium concentration (Equation 5.35) may be divided into two concentrations: C_1, corresponding to sorption in the matrix region (upward curve similar to sorption in rubbery polymers), and C_2, corresponding to sorption in microvoids (downward curve similar to the Langmuir sorption) (Figure 5.6e).

$$C_{eq} = C_1 + C_2 = C_p \frac{k'a}{1 - k'a} + C_p \frac{(A' - 1)k'a}{1 + (A' - 1)k'a} \qquad (5.35)$$

Where:

C_p = weighted mean value of the polymer sorption capacity (a constant value at a given temperature)

This value mainly depends on the polymer structure and its state. k' is the difference between the interactions of vapor molecules with each other and with a polymer molecule segment. High k' values indicate preferential interactions between penetrant molecules compared with the penetrant–polymer interactions; thus, the isotherm shows a stronger exponential increase at high water activity (as clustering formation) when the amount of the new-sorbed water molecules is only due to the interactions between water molecules and the already absorbed water molecules. A' is the difference between the interactions of a microvoid with the molecules sorbed on the first layer and the interactions of a microvoid with the molecules sorbed in the subsequent layers. This value indicates the material capacity to absorb the penetrant molecules after the specific sorption on the Langmuir sites.

This model was successfully applied to the water sorption isotherms of sulfonated or nonsulfonated composite polysulfones with sulfonic-modified Laponite clay and sulfonated poly(ether ether ketone) with sulfonated polyethersulfone Cardo [28].

Park model

Another well-known model, which suits well a sigmoidal shape, is the Park model (Figure 5.6e). This model has five adjustable parameters and contains three sorption modes, i. e. the dual-mode (Henry and Langmuir) combined with the clustering formation:

$$C_{\text{eq}} = \frac{C_{\text{L}} b_{\text{L}} a}{1 + b_{\text{L}} a} + k_{\text{D}} a + n K_{\text{c}} K_{\text{d}}^{n} a^{n} \tag{5.36}$$

This multimode sorption model is well-adapted to describe the water sorption capacity of hydrophilic polymers, bio-based polyesters, and cellulosic compounds such as vegetable fibres.

5.2.2.2 Hysteresis

The sorption hysteresis occurs when the penetrant sorption and the subsequent desorption cycles do not superimpose. The sorption–desorption hysteresis is a well-known phenomenon observed in a wide variety of solvent–polymer systems, but especially in polymer materials, which can swell during sorption (such as hydrophilic polymers, including bio-based polymers, cellulosic materials [29] and proteins, during water sorption). In general, the amount of absorbed molecules at equilibrium state at a constant relative pressure is higher during desorption than during adsorption. To date, there is still no consensus regarding the mechanism of the sorption hysteresis. Several interpretations of this phenomenon can be found in the literature.

In the case of the polymer sorption-induced swelling, there are more polar sites available for the water interaction during the desorption than during the adsorption at the same water vapor pressure. According to White and Eyring [30], an increase in temperature reduces the water adsorption (exothermic reaction), but at the highest relative humidities, the number of adsorption sites in the swollen material increases. It is evident that this number is limited and that a maximum is observed. However, the absence of the maximum means that the mechanical properties of the material play a key role at lower relative humidity. Indeed, as demonstrated by Giacobazzi et al. [31], the moisture content in amorphous and semicrystalline poly(ethylene vanillate) (PEV) films measured during the sorption–desorption cycle is characterized by the hysteresis (Figure 5.7). In addition, this hysteresis is much more pronounced for the amorphous film compared with the semicrystalline film owing to the higher swelling of the amorphous polymer. Thus, the non-Fickian relaxations resulting from the morphological changes in the polymer structure are induced by the swelling–plasticization phenomenon. The water molecules force the molecular chains to move by a swelling effect, so that the secondary bonds between polymer macromolecular chains are broken, thus increasing the access to polar sites. Thus, at the highest possible relative humidity ($a = 1$), the adsorption is limited by the available sites.

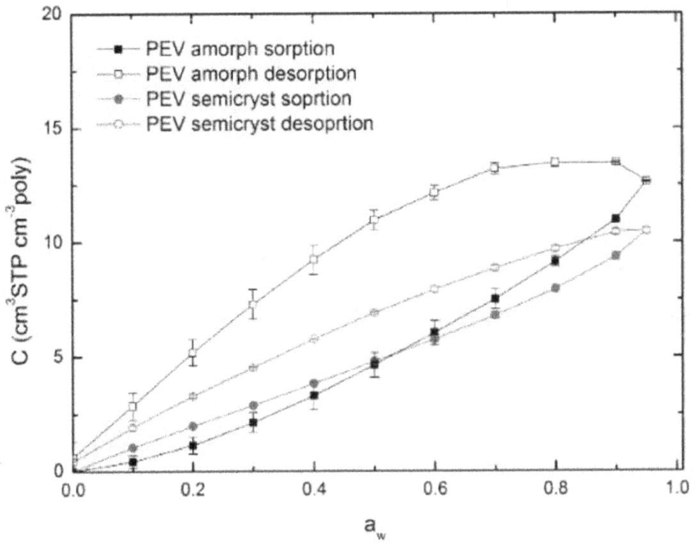

Figure 5.7 Water sorption–desorption hysteresis in amorphous and semicrystalline PEV films at 20 °C. Adapted from [31]

During desorption, the contractive forces of the polymer network are not sufficient to remove adsorbed water molecules. This explanation is also confirmed by Chen et al. [32] in the case of water sorption–desorption hysteresis in nanoporous polymers. Different hydrogen bond networks are analyzed; cellulose swells to form water–cellulose bonds upon adsorption, but these bonds do not break upon desorption at the same chemical potential. From a microscopic model of water sorption in cellulose, cellulose-cellulose hydrogen bonds are reduced in the wet state during swelling and different water-cellulose and water-water hydrogen bond distribution can be formed at the same moisture content during adsorption and desorption. During adsorption, the swelling effect causes break of cellulose-cellulose hydrogen bonds, so that more hydroxyl sites are available to form water-cellulose hydrogen bonds. During desorption, the strong energy associated with cellulose-water hydrogen bonds prevents the cellulose-cellulose hydrogen bonds to reform at the same water activity as during adsorption. Therefore, during adsorption and desorption at a given moisture content, the system will possess the same number of water molecules that will be distributed differently.

On the other hand, it is suggested [33] that the hysteresis is linked to the experimental procedure. It is shown [34] that the hysteresis phenomenon depends on the material softening during sorption in the case of water sorption in cellulosic materials. It has been shown that the sorption–desorption hysteresis decreases with increasing temperature (Figure 5.8) and with increasing softening point. The hysteresis is associated with an energetically unstable structure because of the hardening of the cellulose polymer as the material passes from above T_g to below T_g.

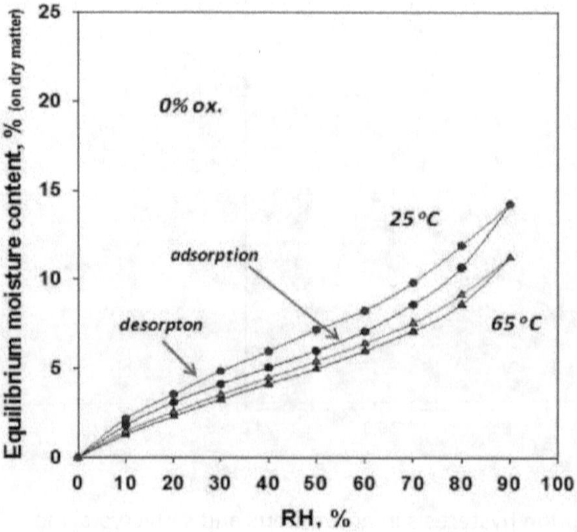

Figure 5.8 Water sorption–desorption hysteresis in cellulose at 25 °C and 65 °C [34]

The hysteresis phenomenon in water sorption–desorption isotherms was also studied for Nafion®, sulfonated polysulfones (SPSU) composites, sulfonated copolyimide (SPI), and sulfonated poly(ether ether ketone) (SPEEK) ionomers [28, 35–38] (Figure 5.9).

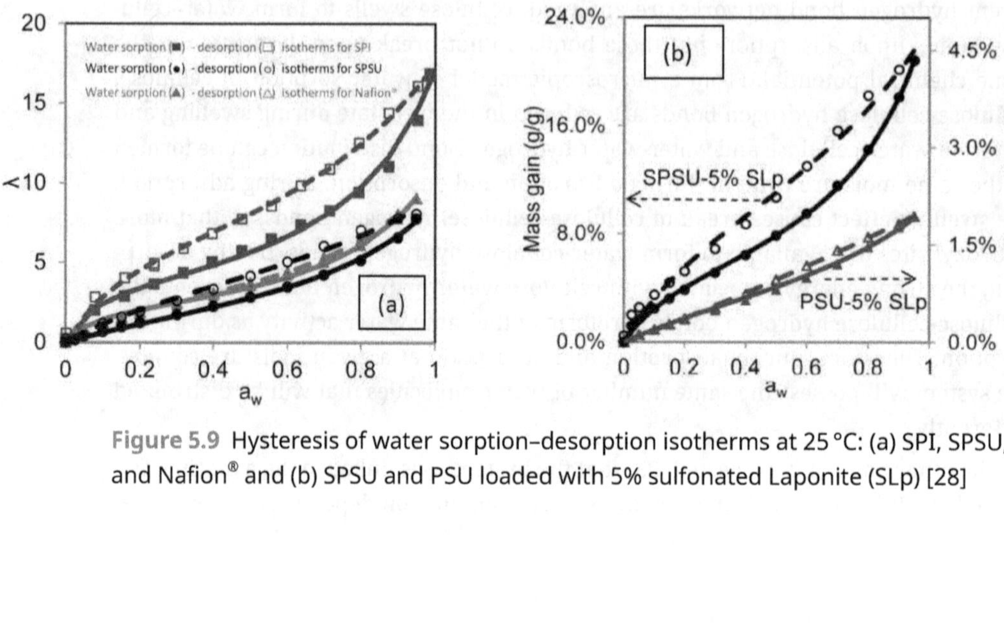

Figure 5.9 Hysteresis of water sorption–desorption isotherms at 25 °C: (a) SPI, SPSU, and Nafion® and (b) SPSU and PSU loaded with 5% sulfonated Laponite (SLp) [28]

Practically no hysteresis can be noted in the case of Nafion® during the sorption–desorption experiments (Figure 5.9a). This result was explained by rather small swelling stress (i. e. chemical potential shift) with small increments in water activity. This stress is so small that the relatively flexible fluorocarbon chains can relax reversibly. In contrast, significant hysteresis can occur when the water activity is changed with large increments. Thus, the large induced swelling stresses can lead to relaxation of longer segments chains, which need more time [39].

Given that hysteresis is expected for hydrophilic polymers (such as SPSU and SPI) due to their significant swelling caused by the high affinity of sulfonated groups with water, it is not surprising to see no sorption hysteresis in the case of nonsulfonated hydrophobic PSU polymers (Figure 5.9b) [28]. Here again, the mass uptake during desorption is higher than that during sorption over the whole range of water activity. Even when the polymer backbone is rather rigid, the water sorption–desorption hysteresis can be large, as in the case of sulfonated polyimides with high T_g values.

To date, the hysteresis sorption–desorption phenomenon is still subject to discussion. However, as suggested by Berens [40] and Wessling et al. [41], the increased free volume resulting from swelling/relaxation phenomena may explain such difference in sorption–desorption cycles. New free volume sites are created and subsequently filled with water molecules during sorption. Rivin et al. [35] proposed the slow volume relaxation of the hydrophilic clusters, when the water vapor activity changes as a result of the membrane swelling. It means that desorption of the already-absorbed water molecules occurs more rapidly than the collapse of the free volume, leaving more free volume available for water molecules and ensuring a higher water vapor concentration during desorption at a given water activity.

5.3 Experimental Aspects

5.3.1 Sorption

Many studies on the water sorption kinetics in polymers can be found in the literature. In general, the mass uptake of a film immersed in liquid water or placed in a humid atmosphere is followed at constant temperature. Sample plates of 1 or 2 g with a thickness of 1–5 mm are used. The measurement extends over several weeks or even several months. The mass increase is generally determined by a sensitive balance (sensitive to changes of 0.1 or 0.2 mg). The measurement uncertainty depends on many factors related to the polymer degradation, to the progress of the operating mode, etc.

A discontinuous method is mostly often used: the sample is introduced into the bath or the enclosed sorption cell, then it is removed after a determined time. It is wiped, if necessary, with absorbent paper, and placed in a closed and tared vessel to carry out the weighing. All these operations are repeated several times until the saturation level is reached. The major drawback of this method is the interruption of the sorption process during the weighing. This limitation can be overcome if the sample mass can be determined directly in the sorption medium. This can be done in the case of humidity-controlled kinetics, carried out below the saturated vapor pressure to avoid water condensation on the film. For example, a McBain balance [2] (Figure 5.10) placed in the sorption chamber can be used. In this case, the mass gain of the sample is linked to the elongation of the quartz spring on which the sample is suspended (for example elongation of 1 mm for 0.1 g with a maximum of 100 mm). However, this simple technique cannot be applied to samples with the absorption capacity of less than 1% by mass. Therefore, microbalances of very high resolution are required for both dynamic and static sorption measurements in the case of barrier materials.

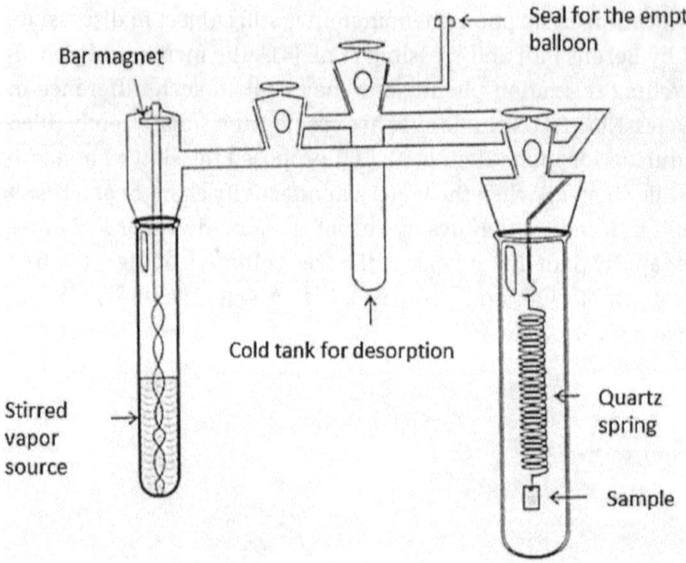

Figure 5.10 Scheme of McBain's balance (quartz spring)

5.3.1.1 Microgravimetry

The microbalance is an ultra-sensitive device allowing high-precision absorption measurements under controlled relative humidity and temperature conditions. Some such microbalances are used for water and organic vapors, but also gas sorption measurements and for adsorption–desorption cycles.

The Sartorius 4201 microbalance was one of the first microbalances [22, 42] used for measuring water or organic vapor sorption kinetics at different relative humidities. This free magnetic suspension balance has a resolution of 2–10 μg and an accuracy of

20 µg (Figure 5.11). This device makes it possible to monitor continuously, by levitation, the weight gain of an isolated sample in its sorption chamber.

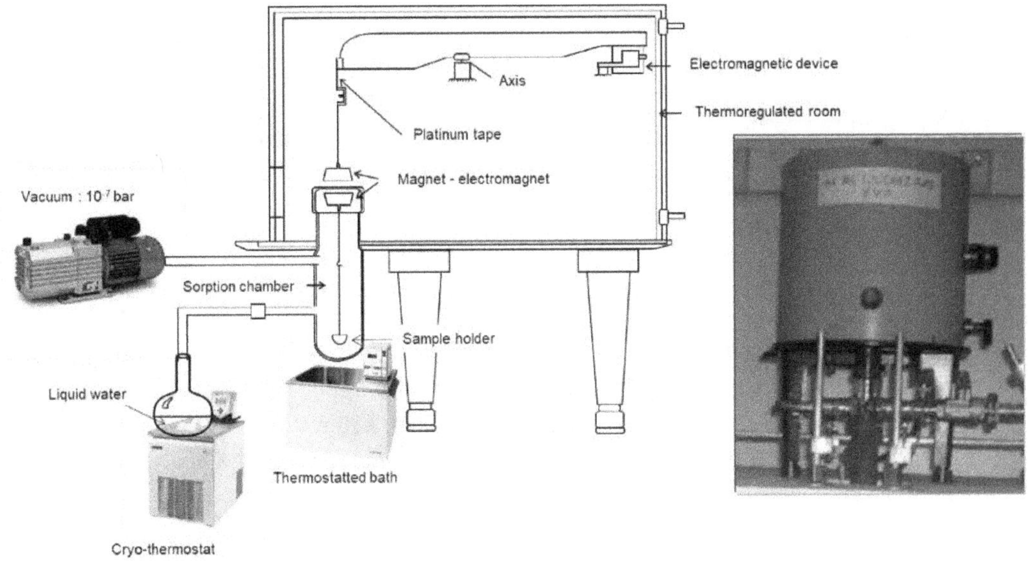

Figure 5.11 Scheme of the Sartorius 4201 microbalance (electromagnetic suspension)

Kontron Instruments has developed an automatic water sorption and desorption measuring device, integrating a Cahn ultra-microbalance installed in an incubator and controlled by humidity, with a dynamic range of 1.5 g with a sensitivity of 0.1 µg and 100 g for a sensitivity of 10 µg. Moylan et al. [43] use a quartz crystal microbalance on which a sample of small dimension is poured. The mass gain is highlighted by a variation in the quartz oscillation. Pranjoto and Denton [44] performed their measurements on a Cahn-1000 microbalance displaying a resolution of 1 µg.

Another type of balance, called the IGA (Intelligent Gravimetric Analyser), from Hiden Analytical Ltd (UK), allows us to perform the measurements with the same resolution [45]. The ultrasensitive microbalance IGA (Figure 5.12) is rated at ±0.1 µg for 10 mg samples and ±1 µg for 5 g samples. This balance can be used in two modes:

1. Static mode with a vacuum purge step (10^{-8} mbar) that allows gas and vapor sorption under pressure (closed system)
2. Dynamic mode with a purge step using dry nitrogen, allowing sorption of water and organic vapors (open system).

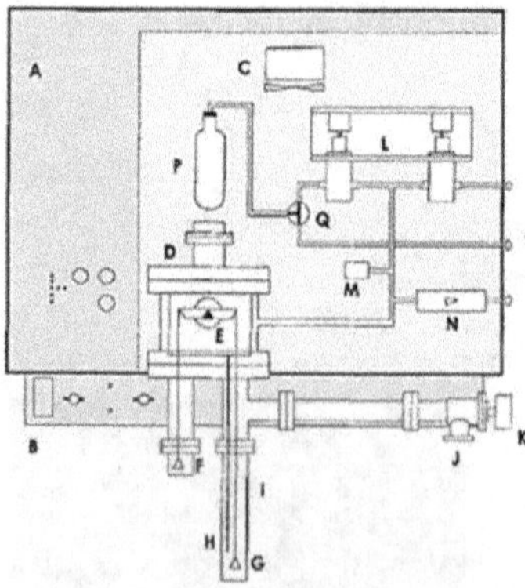

A Thermostated Enclosure
B External Electrical Connections
C Thermoregulator
D Vacuum-Pressure Assemble
E Microbalance Movement (long-term stbility to ± 1μg)
F Tare span
G Sample span
H Temperature Sensors
I Adjustable Baffles
J Vacuum Port (High conductance ports for outgassing)
K Vacuum-Pressure Isolation Valve
L Pressure Controller / Gas Feed
M Pressure Transducer(s)
N Pressure Relief Valve
O External Ports (Gas Admit / Exhaust / Vent)
P Vapor Generator
Q Three Positions Valve

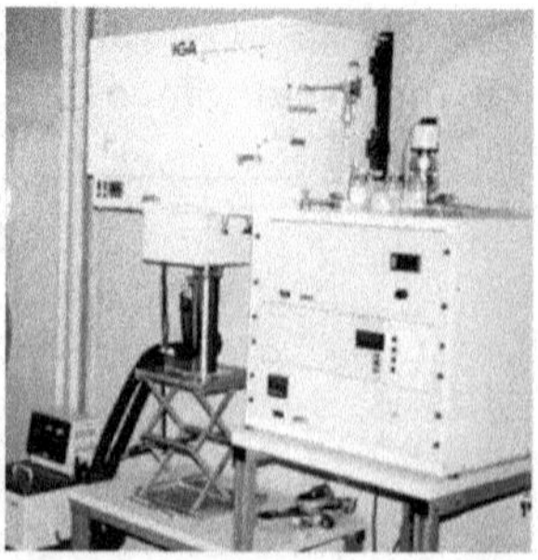

Figure 5.12 IGA microbalance from Hiden Analytical Ltd

A second type of IGA microbalance – the IGAsorp, from Hiden Isochema Ltd (UK), was developed for vapor sorption as well as for water vapor permeation measurements [46]. In this case, the permeation transfer through a film sample is followed by microgravimetry with 0.1 μg resolution for 200 mg sample by using a specific cupel (Figure 5.13).

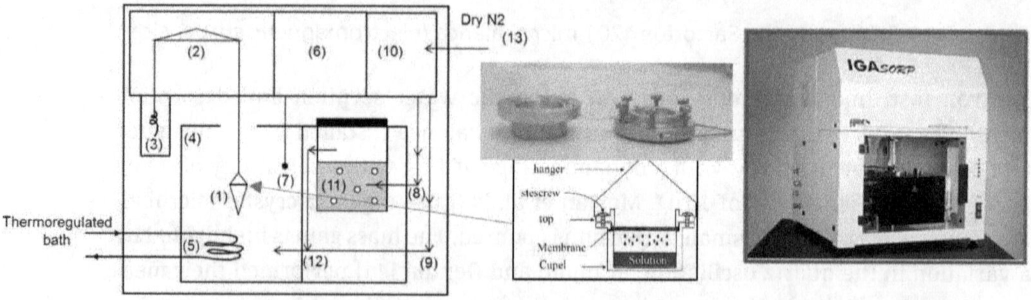

Figure 5.13 Scheme of the IGAsorp microbalance from Hiden Isochema Ltd, used for vapor sorption and vapor permeation measurements: (1) sample holder, (2) microbalance, (3) counterweight, (4) oven, (5) thermocouple, (6) box, (7) relative humidity sensor, (8) and (9) mass flowmeter, (10) box, (11) reservoir, (12) tube with humidified gas, (13) tube with dry gas

The automatic gravimetric vapor sorption DVS (Dynamic Vapor Sorption) instrument, from Surface Measurement Systems (UK) is equipped with one of the most sensitive microbalances (a Cahn D200 with a mass resolution of 0.1 μg), and is used for water or organic vapor sorption/desorption measurements [47, 48] (Figure 5.14).

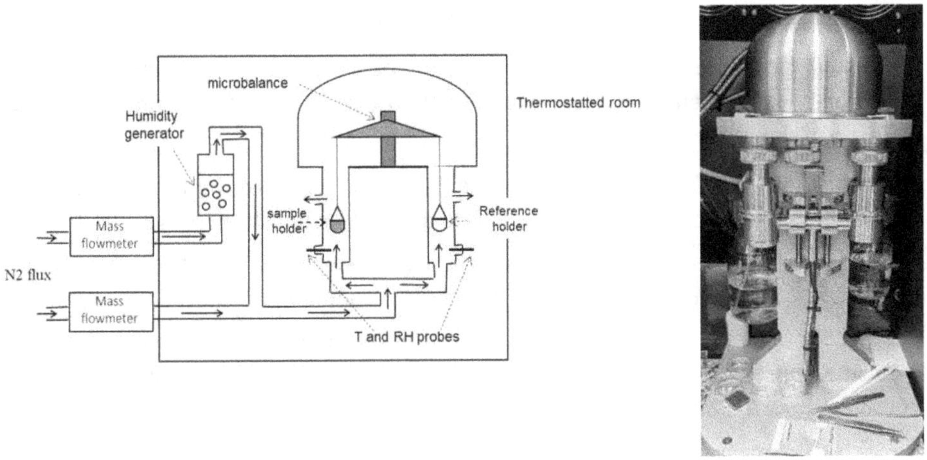

Figure 5.14 The DVS instrument from Surface Measurement Systems

These high-precision microbalances are the most suitable devices for the analysis of polymer sorption capacity in controlled environments. In addition to their ability to measure the vapor sorption in organic/inorganic materials of various forms (i. e. film, powder, liquid), some of them can be combined with different peripheral devices, such as microscopes and infrared spectrometers. Their main limitation is the time necessary to establish a stable weight at a given relative humidity.

5.3.1.2 Data Analysis

Microbalances ensure continuous and precise monitoring of the sample mass gain in the presence of a vapor/gas, for which activity/pressure is well-defined and is imposed in increasing or decreasing order (i. e. sorption or desorption, respectively) and at constant temperature. In the case of water sorption, the water activity is often given by the relative humidity (RH). Figure 5.15 shows the sorbent molecule distribution within the sample exposed to an environment with higher solute activity.

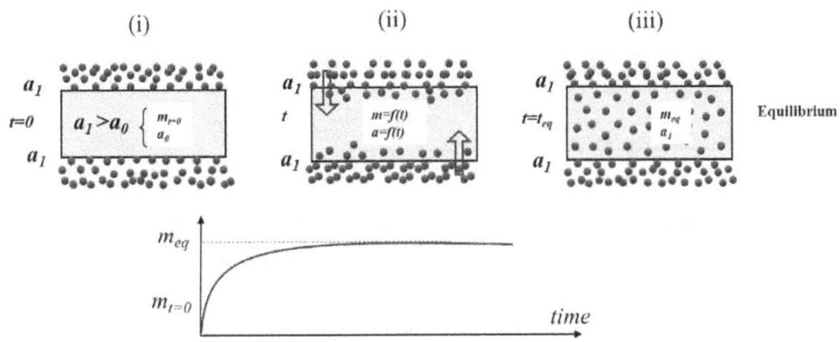

Figure 5.15 Distribution of sorbent molecules during vapor sorption: (i) initially at $t = 0$; (ii) at time $0 < t < \infty$; and (iii) at equilibrium $t = \infty$

As can be seen in Figure 5.15, the molecule penetration takes place in two ways:

- The entire sample surface is exposed to a vapor with an activity a_i and, therefore, the activity gradient is present between the interface and the sample bulk (the plane of the film half-thickness ($L/2$))
- The activity gradient decreases (in intensity) over time from a maximum value ($\Delta a = a_1 - a_0$) to a zero value ($\Delta a = a_1 - a_1 = 0$) at equilibrium state ($t = \infty$).

The sorption rate gradually decreases until a constant mass is reached at the new equilibrium, at which point the sorption process is completed.

The sample (of fixed dimensions) is placed in a sample holder, which is suspended from one end of the microbalance, and the other end is held by a counterweight (stainless steel) to balance the microbalance. The sample is thus located in a temperature-controlled chamber with controlled relative humidity (or vapor pressure) via a probe (dew-point temperature). The temperature is previously fixed at a constant value (isotherm condition). The humidity (from 0 to 95%) is generated by mixing dry and saturated vapor gas flows in the correct proportions using electronic gas flowmeters.

Depending on the type of microbalance used, the sample is initially purged by either a nitrogen dry flow or by a high vacuum. During this step, the sample is dried and all traces of water are removed. The sample mass drops to a constant value, which corresponds to the initial dry mass m_0. After the purge step, the water vapor is introduced at a controlled pressure and the mass uptake is measured as a function of time. A sequence (in absorption mode) is then predefined with increasing activity level (relative humidity) (Figure 5.16).

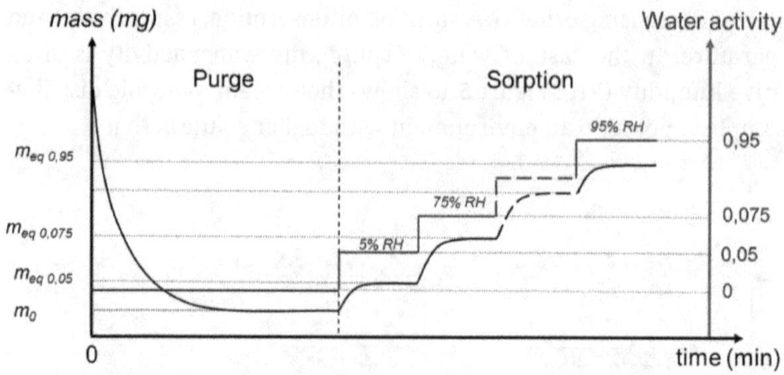

Figure 5.16 Example of a sample mass variation obtained during the purge step followed by water vapor sorption kinetics at different relative humidity

The sorption equilibrium for each activity or vapor pressure is considered to be reached when the mass change in time is below a predefined threshold value. At pressure close to the saturation point, special attention is paid to avoid vapor condensation. The water mass gain at sorption equilibrium (expressed in g of water per g of sample), ΔM_{eq}, is determined at the applied water vapor activity using the following equation:

$$\Delta M_{eq} = \frac{m_{eq} - m_0}{m_0} \tag{5.37}$$

Where:

m_{eq} = sample mass at the equilibrium state

m_0 = sample initial mass (i. e. before absorption)

The sample water sorption behavior is analyzed from kinetic and thermodynamic points of view by means of the methodology described previously in Section 5.2.1 and Section 5.2.2. The accuracy of mass gain values is estimated to be better than 3%. To have better insight in the sorption mechanism, it is essential to correlate the kinetic data with the sorption isotherms. Indeed, depending on the shape of the sorption isotherm, the relevance of the mechanistic model used to describe the sorption phenomenon can be supported by the variation of the diffusion coefficient as a function of the water content, as shown in Figure 5.17.

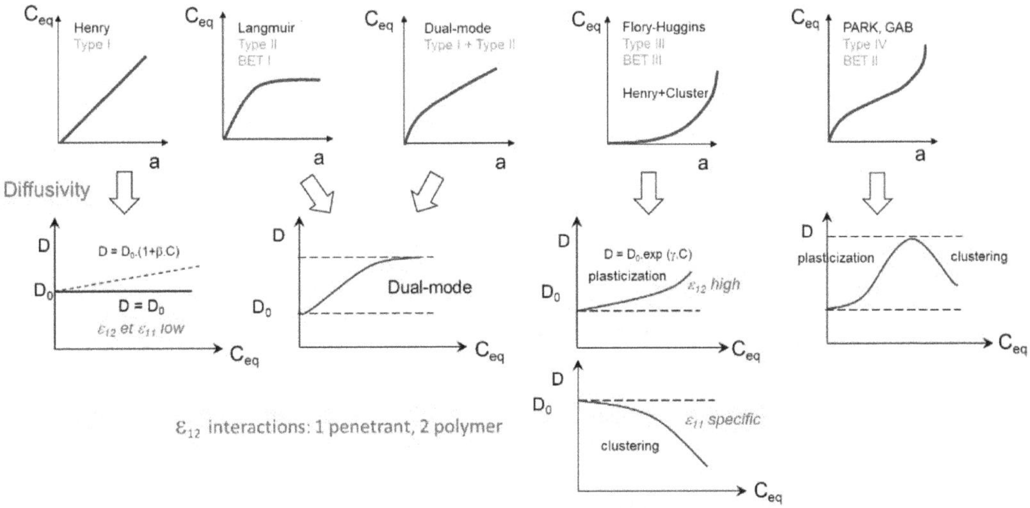

Figure 5.17 Correlation between sorption isotherm and diffusivity as a function of water concentration

5.3.2 Water Sorption Phenomena in Polymers

Due to water–water and water–polymer chain interactions, the water sorption behavior of a polymer material will depend on its hydrophobic/hydrophilic nature, its amorphous or semicrystalline structure, its morphology, and its thermal behavior (i.e. rubbery/glassy state). Due to such interactions and possible water-induced changes of the morphology and structure during sorption, various situations can occur with, in some cases, antagonist effects in diffusivity (positive and negative dependence of D with C) and also a dual dependence on concentration and time due to the relaxation phenomenon induced by the swelling effect. Therefore, the water sorption in polymers is a complex phenomenon, with each water-polymer system being unique, meaning that it should be studied under controlled conditions of temperature and humidity. In addition, these conditions must be very close to the conditions under which the real material is used. From a kinetic point of view and as reported by Simar [49], deviation from the Fickian behavior leads to various kinetic curve shapes owing to the different chemical and physical processes (Figure 5.18).

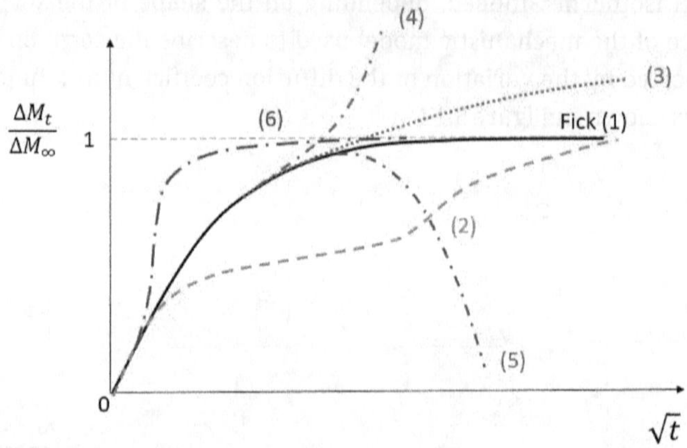

(1) Fickian sorption
(2) Fickian sorption associated with reversible reaction or relaxation process or a coupling with diffusion facilitated by mechanical state
(3) Langmuir model or Fickian sorption + chemical reaction
(4) Fickian sorption followed by degradation and damage of the polymer
(5) Fickian sorption and irreversible chemical reactions (hydrolysis, leaching of materials..)
(6) Fickian sorption and swelling effect due to plasticization / relaxation

Figure 5.18 Examples of Fickian and non-Fickian water sorption kinetics in polymers [49]

5.3.2.1 Plasticization Effect

The plasticization phenomenon is one of the most well-known effect induced by water in polymers and especially in polymers with polar groups that are able to interact with water. Water diffusion through polymers can be the cause of reversible phenomena, such as plasticization by modification of local stresses, but also can be the cause of irreversible phenomena, namely chain breaking and chemical degradation [50]. These two phenomena contribute significantly to changes in the aging kinetics of the material. Plasticization by water absorption may be observed quite easily by differential scanning calorimetry (DSC) through a reduction in the T_g. This reduction is proportional to the quantity of absorbed water [51]. Domack and Johannsmann [52] study the plasticization of thin polymer films by analyzing the impedance of a quartz resonator and by varying the film thickness using an ellipsometer. Marais et al. [53] found a decrease in the T_g of postcured polyester resin after immersion in water. This phenomenon was also observed by Eisenberg et al. [54] in the case of unsaturated polyester resin with different molar mass M_n and styrene/polyester ratio. The authors observed an increase in the T_g for a styrene/polyester ratio close to the value of pure polystyrene, and a decrease in the T_g when the samples are placed at 100% relative humidity and 23 °C for 4 days.

Polyamides (PAs) are well-known to be plasticized by water and alcohols. Ríos de Anda [55] found that the higher the uptake of water, ethanol, and toluene, the more significant is the decrease of the T_g of PA6,6. Also, polar and non-polar solvents have different plasticization trends, as revealed by the DSC measurements, so that for a given molecular uptake, toluene has the highest plasticization effect (i. e. the highest induced T_g drop), followed by alcohols and then water (Figure 2.11).

The influence of water can be explained by the changes caused by the presence of water molecules between the segments of the macromolecular chains and, thus, by a reduction in the interchain interactions. The glass transition (which results from the changes of vibrational motion to micro-Brownian motion with increasing temperature) requires less energy and takes place at a lower temperature as the movement of the chains becomes easier. The comparison of the T_g obtained for an unsaturated polyester resin with the T_g obtained for a linear polymer, such as polyethylene terephthalate (PET) ($\Delta T_g \approx 10$ °C [51]) shows that water plays an important role in the plasticization of the polymer material.

The "free volume" models based on the theory of Cohen and Turnbull [56], i. e. on the relationship between the diffusion coefficient and the material free volume, describe the solvent transport in a polymer matrix by the movement of hard spheres by jumps in the matrix micro-voids. These models take into consideration all the factors that can influence the free volume formation, the glassy or rubbery character, as well as the specific size of the polymer micro-voids. The movement of a diffusing molecule is controlled over time by a redistribution of voids (cavities) around the molecule due to thermal fluctuations. According to Cohen and Turnbull, the entire diffusing molecule is moving without considering the mass effect [56].

Another version of the free volume model is that of Vrentas et al. [57]. In this case, the energy required for the penetrant movement and the increase in free volume resulting from the polymer thermal expansion are taken into account. The Vrentas–Duda free volume model characterizes the solvent diffusivity in a polymer in terms of solvent size and concentration, and the thermal distance between temperature and the T_g of each component (i.e. penetrant and polymer). Also, among all models of free volume, the Fujita model expresses D according to the fraction of free volume V_f as follows [58]:

$$D = RTA_f \exp\left(\frac{-B}{V_f}\right) \tag{5.38}$$

Where:

A_f and B = characteristic parameters, which depend on the nature of the penetrant–polymer system and increase rapidly with the permeant size

V_f = free volume fraction, a linear function of the concentration:

$$V_f = V_{f2} + (V_{f1} - V_{f2})\phi \tag{5.39}$$

Where:

V_{f1} and V_{f2} = fractions of the free volume of the penetrant and the polymer (respectively)

ϕ = fraction of the penetrant in the polymer (~C)

For low penetrant concentrations, taking $D_0 \approx \exp(-B/V_{f2})$ and $\gamma \approx B(V_{f1} - V_{f2})/V_{f2}^2$, Equation 5.38 becomes Equation 5.18.

In Equation 5.18, the γ parameter is a plasticization coefficient. The physical meaning of the γ coefficient is not totally clear. Kokes and Long [59] suggested that this parameter is proportional to the Flory–Huggins interaction parameter χ. The γ parameter is related to the increase of the polymer chain segment mobility induced by the presence of the penetrant. Indeed, the minimum void polymer volume required to disperse the penetrating molecules is not negligible compared with the average free volume, so that the chain segments should rearrange themselves, allowing the diffusive species to jump from one cavity to another. According to Frisch [4], this fact corresponds to the Fickian type B diffusion mechanism generally observed in the case of large organic vapor molecules, and is characterized by: (1) an undiluted polymer–penetrant mixture; (2) a marked dependence of the diffusion coefficient on the concentration; and (3) a temperature dependence of the activation energy.

The water plasticization effect in crosslinked polyester resins has been shown from water sorption measurements performed at 25 °C [60, 61]. The water vapor sorption kinetics measured at different relative humidities allow us to plot the water sorption isotherm (Figure 5.19) and to determine the diffusion coefficient of the first-half D_1 (Equation 5.6) and second-half D_2 (Equation 5.7) sorption as a function of the water concentration C_{eq}.

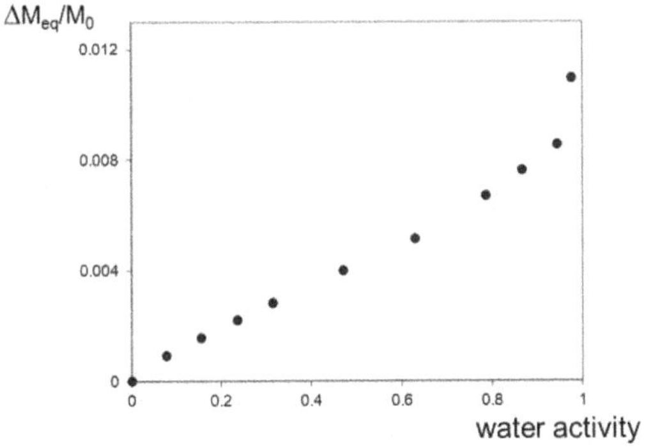

Figure 5.19
Water sorption isotherm of crosslinked polyester resin at 25 °C

As can be seen from Figure 5.19, the shape of the isotherm curve is of type BET II and can be described by the Park model (Equation 5.36) with the dual-mode sorption (due to the glassy state of the crosslinked material, $T_g \sim 75\,°C$) and water clustering formation.

Higher D_2 values compared with D_1 values testify to the increase of D with C. Also, it has been observed that $\ln D_1$ is a linear function of C_{eq} (Figure 5.20a). This result agrees well with the exponential law of D with C (Equation 5.18). Thus, the experimental data of water sorption kinetics can be correctly fitted with Equation 5.18 (Figure 5.20b) by applying the method developed by Marais et al. [7].

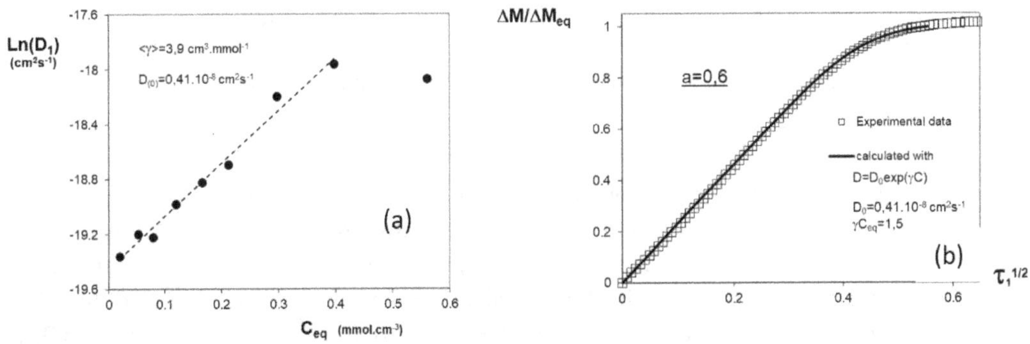

Figure 5.20 Water diffusivity in crosslinked polyester resin at 25 °C: (left) variation of the water diffusion coefficient (ln*D*) as a function of water concentration and (right) experimental water sorption curve fitted with Equation 5.18

The water sorption isotherms of cellulosic materials can be also described by the Park model (Equation 5.36) combining three sorption modes, i. e. Henry, Langmuir and aggregation (Figure 5.21a).

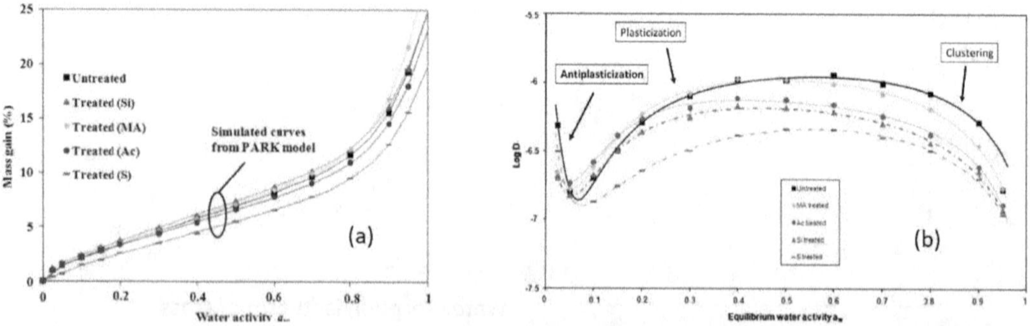

Figure 5.21 Water sorption in cellulosic flax fibers at 25 °C, untreated and treated fibers by chemical treatment with styrene (S), maleic anhydride (MA), acetic anhydride (Ac) or silane (Si): (left) water sorption isotherms fitted by the Park model, and (right) water diffusion coefficients as a function of water activity [62]

In general, the swelling effect caused by the water plasticization phenomenon leads to the water diffusivity increase in the range of relative humidity from 0 to 50% RH (Figure 5.21b). In some cases of cellulosic fibers, namely flax fibers, a slight decrease of water diffusivity at low water content (at 0–0.05% RH) can be observed. This fact is attributed to an antiplasticization phenomenon [62]. Indeed, such results may be explained by the tightening of cellulosic chains induced by the hydrogen bonding between water and hydrophilic sites of cellulosic compounds. This cohesion increase leads to a decrease of the free volume and, thus, a slowing down of the passage of water molecules. Similar results were reported by Chang et al. [63] for tapioca starch films and by Marzec and Lewicki [64] for cereal-based products.

5.3.2.2 Clustering Formation – Zimm–Lunberg Theory

As water molecules can interact with each other, the water clustering phenomenon can occur in polymers whatever the polymer's hydrophobic/hydrophilic nature. Indeed, water clusters are observed in many polymers – for example, in hydrophobic polymers, such as polyolefins (polyethylene, polypropylene) [65], but also in hydrophilic polymers, such as polyamide 6 (PA6) [55, 66–69] and biobased polymers [48], as well as in polyelectrolytes [28].

As previously mentioned in Section 5.2.2.1, the water sorption isotherm with a convex curve at high water activity is typical of the presence of water aggregates (Figure 5.6e). An upward shape curve of water and ethanol sorption isotherms for PA6 films (at 40 °C) plotted as a function of the number of sorbed molecules per monomer unit in the amorphous phase was found by Sabard et al. [70] (Figure 5.22). Two types of PA6 films (PA6/50 and PA6/125) were elaborated by a co-rotating screw extruder with the chill-roll temperature fixed either at 50 °C or at 125 °C.

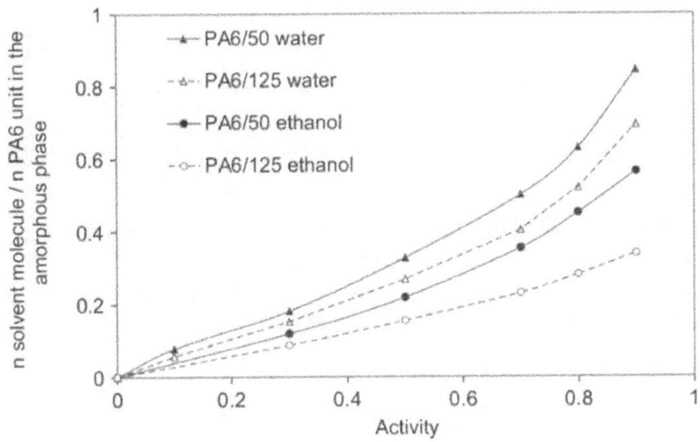

Figure 5.22 Water and ethanol sorption isotherms of PA6 films (at 40 °C) expressed as the number of sorbed molecules per monomer unit in the amorphous phase [70]

In the case of poly(hydroxyalkanoates) (PHA) films (biodegradable polyester), Follain et al. [48] showed that the sigmoidal shape of the water sorption isotherm curve (type BET II) was well-fitted by using the multi-mode sorption of the Park model (Equation 5.36), for which the third part of the curve (i. e. an exponential increase of the water mass gain for highest relative humidity) is relative to the water clustering phenomenon (Figure 5.23).

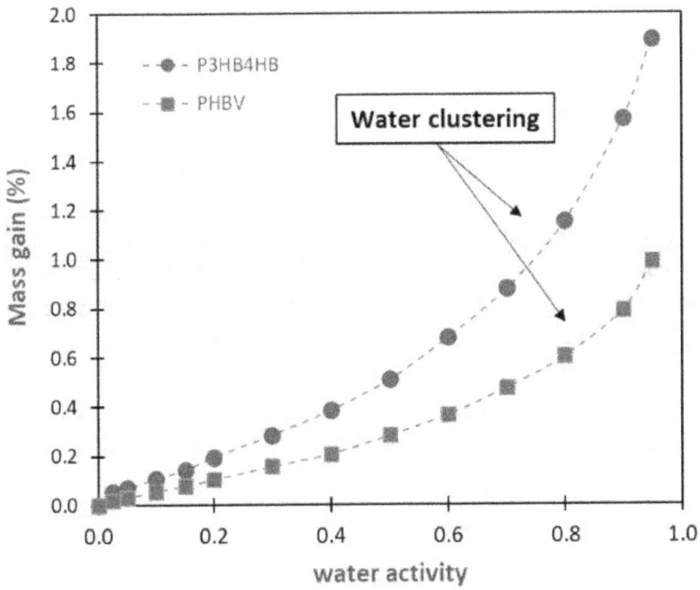

Figure 5.23 Water sorption isotherm of PHA films – poly(3-hydroxybutyrate-*co*-3-hydroxyvalerate) (PHBV) and poly(3-hydroxybutyrate-*co*-4-hydroxybutyrate) (P3HB4HB) [48]

The water clustering formation is known as a relative slow kinetic process, which occurs when a certain water concentration is reached, usually over 50% RH. From a kinetic point of view, the water diffusivity decreases due to the water cluster size, which makes the mobility of water molecules inside the polymer more difficult. This decrease of D with C owing to the clusters of water molecules has been already observed in various polymers, such as polyolefins [65], biopolyesters (PHA and polylactic acid (PLA)) [48, 71] (Figure 5.19), polysulfones (Figure 5.20), and Nafion® [28]. In the case of sulfonated tetrafluoroethylene-based fluoropolymer-copolymer Nafion®, the water sorption clustering is evidenced by a strong increase of upward curve at high water activity [72], as shown in Figure 5.24.

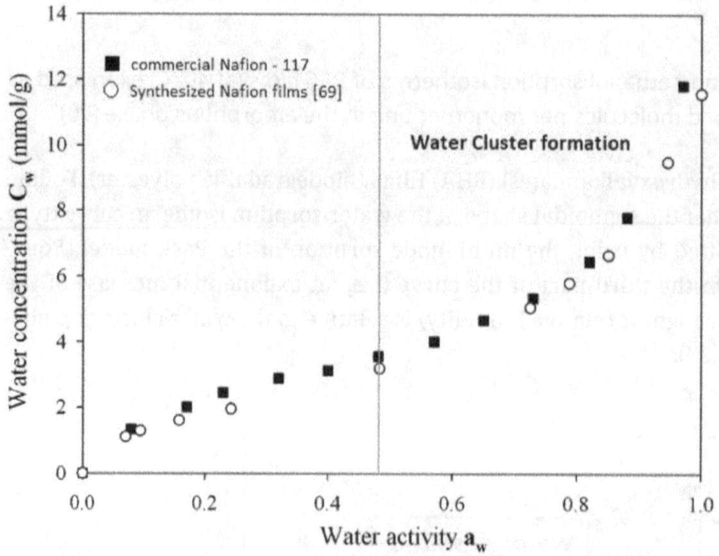

Figure 5.24 Water vapor sorption isotherm of Nafion® films at 25 °C [72]

A decrease of water diffusivity during the water sorption process is expected as a consequence of the hydrogen-bond clustering, which occurs close to sulfonic groups and in the region of the hydrophobic fluorinated backbone [7]. Due to the steric hindrance of water aggregates, the mobility of water clusters is reduced with an increase of the water concentration in Nafion®. The fitting of the water sorption kinetics curve at 95% RH at 25 °C by using the diffusion law (Equation 5.18) but with a negative value of γC_{eq} reveals the decrease of D with C owing to the water cluster formation (Figure 5.25).

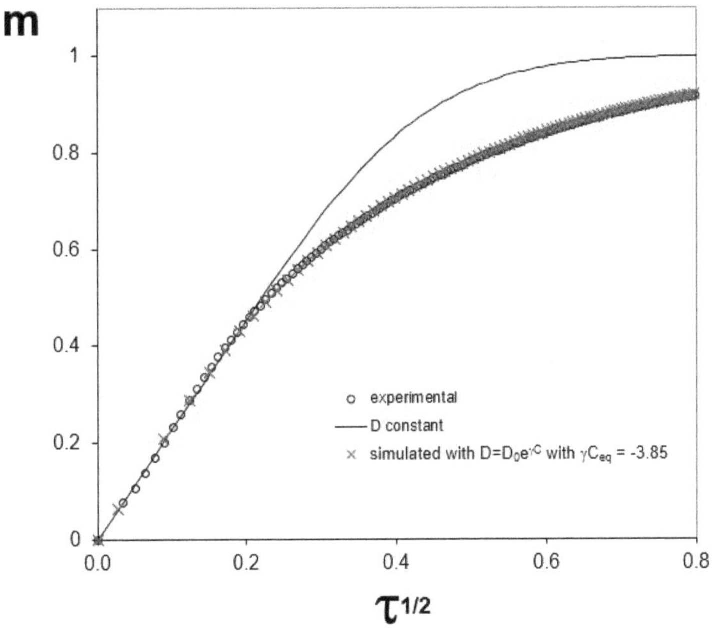

Figure 5.25 Experimental and simulated (Equation 5.18, $\gamma C_{eq} = -3.85$) curves of reduced water uptake of Nafion® film ($L = 170\ \mu m$) (95% RH, 25 °C) as a function of the dimensionless time $\tau^{1/2} = (Dt)^{1/2}/L$

It should be mentioned that water clustering can occur not only in hydrophilic polymers, but also in hydrophobic polymers (for example, polyolefins and polycarbonates [73]), as water–water interactions can easily take place. For example, water vapor permeation fluxes through polyolefin films (polyethylene (PE) and polypropylene (PP)) and poly(ethylene-*co*-vinyl acetate) (EVA) films with low vinyl acetate content have been analyzed [13, 65], and the water diffusivity is found to decrease due to water cluster formation. The water migration decrease is underlined by the negative plasticization effect determined by the curve fitting (Figure 5.26). This clustering formation is much easier when the polymer is in a rubbery state, as high free space allows aggregates to be formed.

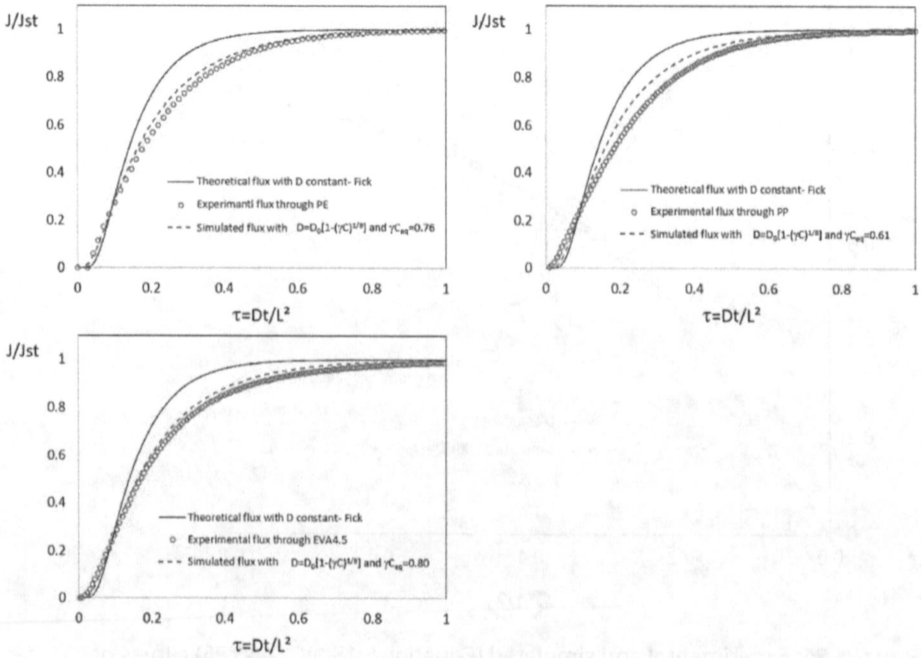

Figure 5.26 Experimental and simulated water flux curves through PE, PP, and EVA in dimensionless scales of flux and time

Zimm–Lunberg theory – Mean cluster size (MCS)

The clustering theory of adsorbed molecules was proposed to explain some of the sorption thermodynamic contradictions first by Zimm et Lundberg [74, 75] and later by Starkweather [76]. Water is a unique penetrant because, due to its polar nature, it can create hydrogen bonds with itself and can form clusters. This theory allows us to evaluate the degree of clustering in binary systems. The clustering function is thus defined as:

$$\frac{G_S}{V_S} = -(1 - \phi_S)\left[\frac{\partial(a/\phi_S)}{\partial a}\right]_{p,T} - 1 \tag{5.40}$$

Where:

G_s = cluster integral

V_s, ϕ_S, a = partial molar volume, the volume fraction and the activity of penetrant molecules (in our case, water molecules) (respectively)

The clustering function G_s/V_s indicates whether clustering takes place or not, and it can be easily determined from the experimental isothermal sorption [77, 78]. A G_s/V_s value equal to –1 (the case of the ideal solution, i. e. with a sorption isotherm following Henry's law (Figure 5.6e)) indicates that penetrant molecules do not affect the distribution of other penetrant molecules. If $G_s/V_s > 0$, penetrant molecules tend to cluster resulting in a higher concentration in their neighborhood, as would be expected on the basis of non-random mixing, whereas if the G_s/V_s value is greater than –1, penetrant molecules prefer to remain isolated.

The expression $G_S\phi_S/V_S$ represents the mean number of penetrant molecules in excess of the mean concentration. Thus, it measures the clustering tendency of penetrant molecules. The MCS value is hence an estimate of the mean number of molecules in a cluster:

$$\mathrm{MCS} = 1 + \left(\frac{\phi_S G_S}{V_S}\right) \tag{5.41}$$

Following Starkweather [76], the equation for the mean size of a cluster can be written as:

$$\mathrm{MCS} = (1 - \phi_S)\left(\frac{\partial \ln \phi_S}{\partial \ln a}\right)_{p,T} \tag{5.42}$$

The volume fraction ϕ_S is obtained from the relation:

$$\phi_S = \left(1 + \frac{\rho_S}{M\rho_p}\right)^{-1} \tag{5.43}$$

Where:

ρ_s and ρ_p = densities of the penetrant (water) and polymer (respectively)

The mass uptake ΔM (= M_s/M_p) at the equilibrium and at a given activity a of sorbed species is defined according to Equation 5.37. In the case of the Henry–clustering mode, the mass uptake can be expressed according to Equation 5.32.

In the case of water transport studies, the Zimm–Lundberg theory has been often coupled with the approaches that allow obtaining a mathematical description of the water sorption isotherms. As a result, the MCS value can be expressed from the various models involving clustering formation (Table 5.1).

Table 5.1 Equations for Determining the MCS Value According to Sorption Models

Sorption model	MCS equation ($\rho = \rho_s/\rho_p$)
Henry–clustering	$\mathrm{MCS}_{\mathrm{Henry-cluster}} = \frac{\rho^2}{M^3(1+\frac{\rho}{M})^2}\left[k_D a + n^2 K_c k_D^n a^n\right]$
GAB	$\mathrm{MCS}_{\mathrm{GAB}} = \frac{\rho^2}{M^2(1+\rho/M)^2}\left[1 - \frac{M}{M_m C}(-2Ka(C_G - 1) - 2 + C_G\right]$
Park	$\mathrm{MCS}_{\mathrm{Park}} = \frac{\rho^2}{M^3(1+\rho/M)^2}\left[k_D a + \frac{C_b a}{(1+ba)^2} + n^2 K_c k_D^n a^n\right]$
NDMS	$\mathrm{MCS}_{\mathrm{NDMS}} = \frac{\rho^2}{M^3(1+\rho/M)^2}\left[C_p\frac{k'a}{(1-k'a)^2} + C_p\frac{(A'-1)k'a}{(1+(A'-1)k'a)^2}\right]$

In constrast to sorption models with a fixed parameter n representing the number of water molecules in a cluster, this theory allows to see that this number varies with the water concentration in the whole water activity range.

The water sorption behavior of polyamides has been studied in detail, especially for PA6 [67, 70, 79, 80] and PA6,6 [69, 76, 81]. The water sorption isotherms are fitted by the GAB model (Equation 5.34) (i. e. the multilayer clustering model) and the MCS values are calculated. It is shown that at water activity close to 1 (i. e. saturated water vapor) the water clusters are statistically composed of ~3 molecules in the case of PA6,6 at room temperature and of ~2–3 molecules in PA6 depending on the temperature (Figure 5.27).

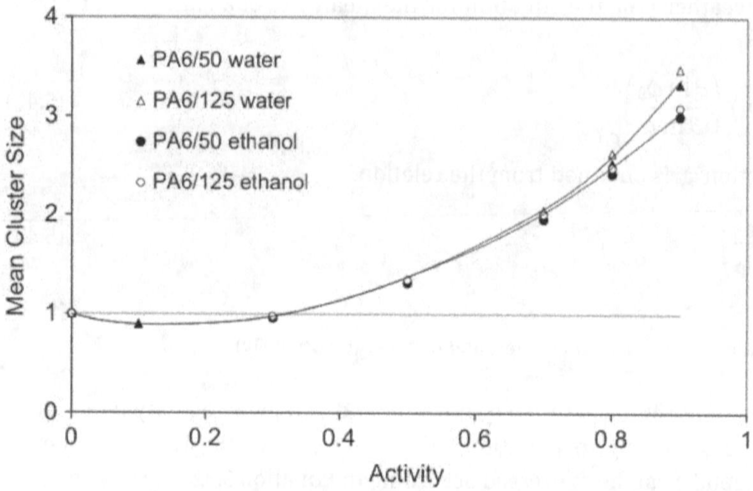

Figure 5.27 Evolution of the MCS of water and ethanol as a function of the solvent activity for PA6 films [70]

The polymer water sorption capacity is found to be higher for polymers with hydrophilic sites, such ionomers, in which the ionic groups fixed on the backbone can be easily solvated by water molecules. The well-known perfluorinated polymer Nafion® (used as a proton exchange membrane for fuel cells) contains sulfonic $-SO_3H$ groups, which strongly interact with water molecules [82]. It is assumed that an ion-conducting channel is formed between sulfonic groups during the water sorption depending on the membrane morphology. In the literature, some models (namely, cluster–channel, core–shell, rod, sandwich, and water channel models) are used to describe the sorption and conductive properties of Nafion® membranes [83]. The cluster–channel morphology model considers the Nafion® sulfonic groups to be arranged in the form of clusters (diameter ~4–6 nm) that are interconnected through narrow channels (diameter ~1 nm) (Figure 5.28a) [84, 85], while in the more recent water channel model, the sulfonic groups are self-assembled into blocks of hydrophilic water channels (each ~2.5 nm in diameter), which ensure ion transport [83, 86]. According to Kreuer et al. [87], the Nafion® structure in the presence of water can be considered as a nano-separation between hydrophobic and hydrophilic regions, as illustrated in Figure 5.28b.

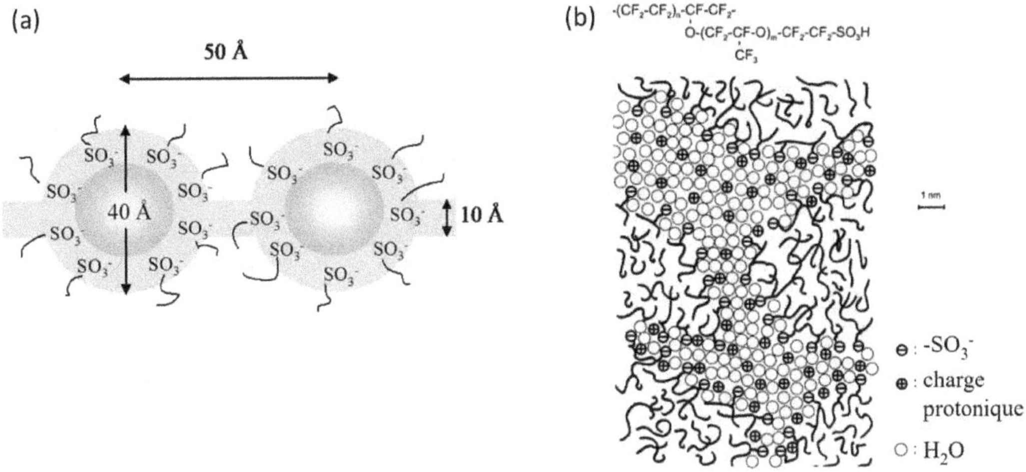

Figure 5.28 Structure of Nafion® according to (left) Gierke et al. [84] and (right) Kreuer et al. [87]

It is revealed that the morphology of Nafion® changes strongly with the water content, so that in the swollen state, water channels are present, allowing proton transfer through the migration of water aggregates that initiate the vehicle diffusion mechanism (called "Einstein diffusion") in addition to the Grotthuss mechanism. Li et al. [88] explained the Nafion® water-hydrophilic structure at different water contents at 80 °C by plotting the free volume per water as a function of the number of sorbed molecules per ion pair (Figure 5.29).

The Nafion® water sorption behavior is still discussed, especially when the water vapor saturation is reached due to discontinuity of the water concentration between saturated water vapor and the liquid state. This phenomenon is known as Schroeder's paradox.

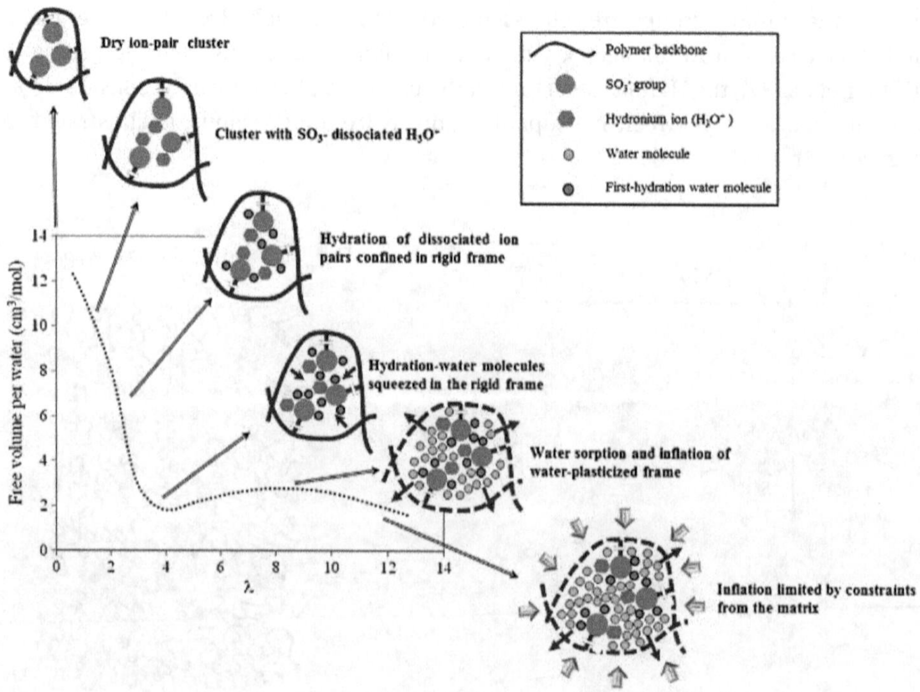

Figure 5.29 Representation of the water-hydrophilic Nafion® structure at 80 °C at different water contents λ (i. e. sorbed water molecules per ion pair) associated with the free volume variation [88, 89]

The presence of water aggregates is also demonstrated in the case of many other sulfonated polymers – for example, in IonClad® [90]. Li et al. [28] determine from the sorption kinetics the water diffusion coefficients of SPSU films loaded with a sulfonic-modified Laponite clay (SLp). The strong decrease of D at high water activity ($a > 0.5$) originates from the water clustering (Figure 5.30a). On the basis of the NDMS model (Equation 5.35), the MCS-1 quantity expressed as the mean number of penetrant molecules in the neighborhood of a given molecule in excess of the penetrant mean concentration in the polymer is calculated. Practically, the same mean cluster size at water activity $a > 0.7$ for SPSU and its composites is obtained (Figure 5.30b), thus highlighting the dominant role of SPSU in the water sorption behavior of SPSU composites.

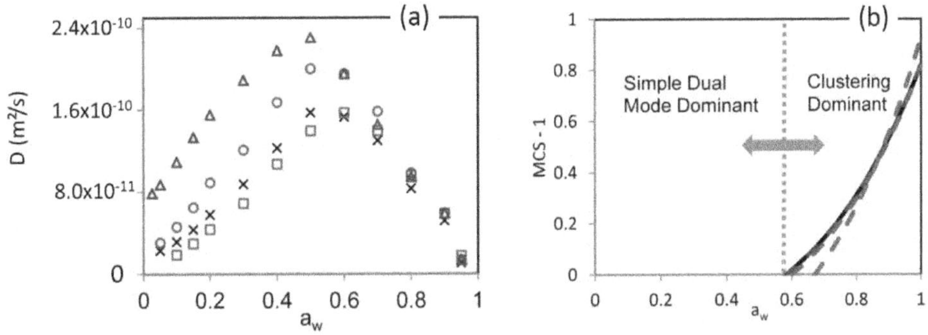

Figure 5.30 (Left) Diffusion coefficient determined from water sorption kinetics for SPSU and SPSU-based composites (SPSU (×), SPSU–1%SLp (□), SPSU–3%SLp (○), and SPSU–5%SLp (△). (Right) Mean cluster size as a function of the water activity [28]

In the case of composite/nanocomposite materials, the presence of inorganic fillers in the polymer matrix can bring new hydrophilic sites for water interactions, and can therefore initiate the formation of additional water clusters. This is the case for poly(hydroxyalkanoates) (PHA)/Montmorillonite films, for which the MCS value is calculated as a function of the water activity [48]. In this case the water sorption behavior of two series of nanocomposites (poly(3-hydroxybutyrate-*co*-3-hydroxyvalerate) (PHBV)/C30B and poly(3-hydroxybutyrate-*co*-4-hydroxybutyrate) (P3HB4HB)/C30B with C30B mass fraction of 2.5 to 10 wt.%) was investigated. Due to their hydrophilic nature, the organo-modified Montmorillonite clay (C30B) has a water sorption capacity higher than that of the hydrophobic polyester matrix – at water activity $a = 0.8$, the mass gain is 4.3% for C30B, 0.6% for PHBV and 1.15% for P3HB4HB. The clay influence on the water cluster formation is studied by analyzing the MCS value deduced by modelling the water sorption isotherms using the Park model (Equation 5.36). It is shown that the water cluster formation as well as the MCS value increase by the nanoclays presence. This tendency is especially observed in PHBV-based films, in which the crystalline phase fraction is higher compared with P3HB4HB-based films (i. e. ~60% versus ~32%) and in which the proportion of clay aggregates is less (Figure 5.31). Thus, the water clustering phenomenon seems to be favoured in confined spaces at the vicinity of hydrophilic particles. More inorganic particles are dispersed into the polymer matrix, more water molecules are aggregated on the hydrophilic surfaces.

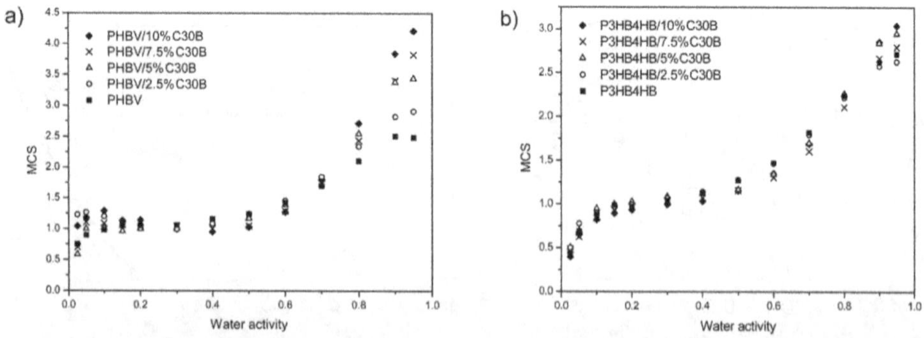

Figure 5.31 Mean cluster size of water as a function of water activity in a) PHBV/C30B and b) P3HB4HB/C30B nanocomposites [48]

5.3.2.3 Relaxation Phenomenon

As previously discussed, the plasticization phenomenon occurs when solvent-polymer interactions take place during the sorption process, leading to an increase of the solvent diffusion coefficient (Fickian diffusion of type B). However, in the case of glassy polymers for which the plasticization effect occurs and provokes a decrease of the glass transition temperature close to room temperature and to the temperature of experiment, a change of the transient diffusion regime during the solvent uptake can be observed owing to the local material plasticization by the solvent (see Figure 5.18, curve 6 and Figure 5.32). This case of a type II diffusion is described by Berens and Hopfenberg [91] as a combination of a Fickian regime and a relaxation process.

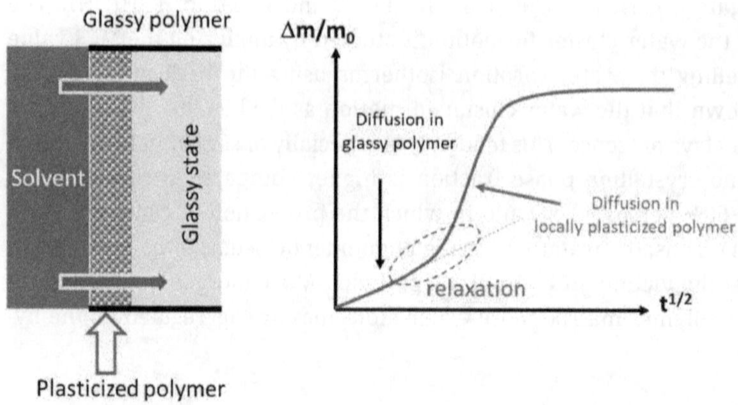

Figure 5.32 Schematic representation of a type II diffusion and mass uptake as a function of $t^{1/2}$ for a relaxation phenomenon during sorption process [55]

The analytical expression of the sorption process in glassy polymers is given by:

$$M_t = M_{t,F} + M_{t,R} = M_{\infty,F}\left(1 - \frac{6}{\pi^2}\sum_{n=1}^{\infty}\frac{1}{n^2}e^{(-n^2K_Ft}\right) + M_{\infty,R}\left(1 - e^{(-Kt)}\right) \tag{5.44}$$

Where:

M_t = total amount of sorption per unit weight of polymer microspheres at time t

$M_{\infty,F}$ and $M_{\infty,R}$ = contribution of the Fickian and relaxation processes, respectively, at time t

The K_F parameter is equal to $4\pi^2D/d^2$, where D is the diffusion coefficient and d is the particle diameter. K is the relaxation-rate constant. This expression allows us to separate the contribution related to diffusion from the contribution related to the first-order relations.

This two-stage model is successfully used to fit sorption kinetics of alcohols [92] in polyethylene terephtalate (PET). According to Berens, the diffusion and plasticization processes occur at the same time, while Ramani and Ranganathaiah [93] assume that the plasticization process would start after the diffusion beginning. In order to simulate water and ethanol sorption kinetics in PA6, Sabard et al. [70] modified the empirical model developed by Berens and Hopfenberger by integrating a delay time t_D and by considering that the relaxation was delayed (Figure 5.33):

$$\frac{m_t}{m_{eq}} = \varphi_{sF}\left(1 - \frac{8}{\pi^2}\sum_{n=0}^{\infty}\frac{1}{(2n+1)^2}e^{\left[\frac{-(2n+1)^2\pi^2Dt}{L^2}\right]}\right) + \varphi_{sR}\left(1 - e^{[-k_r(t-t_D)]}\right) \tag{5.45}$$

Where:

$\varphi_{sF} = M_{eqF}/M_{eq}$, $\varphi_{sR} = M_{eqR}/M_{eq}$ and $\varphi_{sF} + \varphi_{sR} = 1$, M_{eqF} = amount of solvent sorbed at equilibrium according to Fickian diffusion

M_{eqR} = amount of solvent sorbed at equilibrium according to the relaxation process

This modified model was also applied to fit acetone sorption kinetics in PET [94].

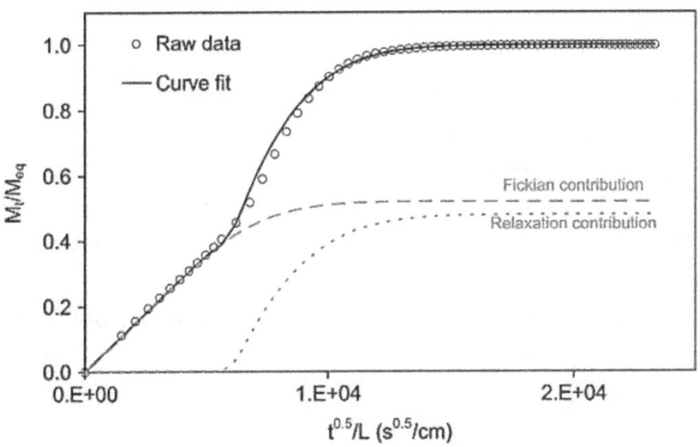

Figure 5.33 Experimental and calculated sorption kinetics curves using the Berens-Hopfenberger diffusion-relaxation model [70]

The relaxation phenomenon induced by sorption during the water diffusion was observed in the sulfonated poly(ether ether ketone) (S-PEEK) glassy polymer [95]. As shown in Figure 5.34a, this relaxation phenomenon (which leads to additional mass uptake after the initial mass uptake) is more pronounced when the water activity is high. Also, the calculated values of the glass transition temperature T_g using the Fox equation decrease significantly as a function of the water concentration inside the polymer (Figure 5.34b). The calculated T_g value of the swollen glassy polymer reaches practically the experimental temperature of 20 °C. This happens in the region of the polymer transition from a glassy state (where both Fickian diffusion and relaxational changes may occur during sorption) to the rubbery state (where only Fickian sorption kinetics takes place and the relaxation contribution is considered to be zero as the polymer is in its fully relaxed state).

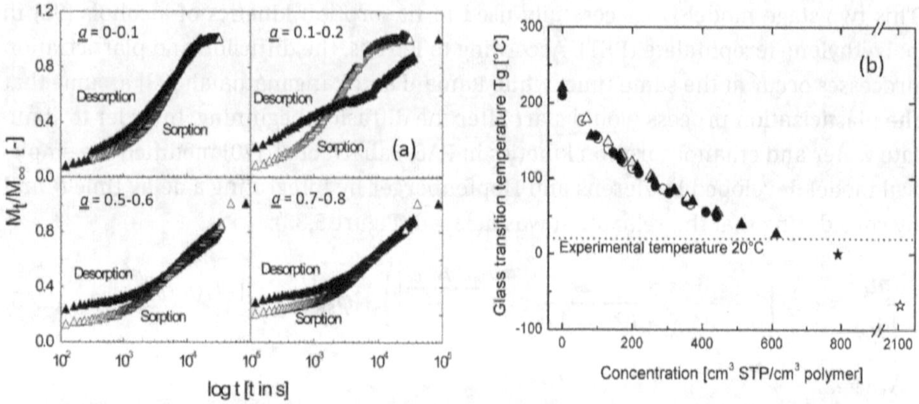

Figure 5.34 (a) Experimental water vapor sorption/desorption kinetics in S-PEEK with a sulfonation degree of 75% at different water vapor activity and (b) the calculated T_g values as a function of the water concertation [95]

5.3.2.4 Water-Induced Crystallization

The water plays a rather complex role in polymers, in particular owing to its specific interactions with the polar sites of the polymer chains, thus changing the material morphology during the sorption process. This is the case of the solvent-induced crystallization phenomenon which leads to the microstructure modifications and, subsequently, to the evolution of the material physicochemical properties, such as mechanical, thermal, optical and barrier properties. The semi-aromatic polyamide poly(*m*-xylene adipamide) (MXD6) widely used in packaging (for example, with PET in multilayer form for blow-molded bottles) due to its barrier properties is known to be characterized by the water-induced crystallization [96]. Indeed, Ben Doudou et al. [97] have shown that initially amorphous MXD6 in contact with water is able to be highly plasticized, its glass transition temperature decreases from 82 °C to the temperature lower than the ambient temperature; the macromolecular chain segment mobility becomes so high (polymer in the rubbery state) that MXD6 can undergo a crystallization by rearrangement of the molecular chains (Figure 5.35). It is shown by the DSC analysis, that the degree of crystallinity decreases gradually from the polymer/water interface to the core and increases when the time immersion in water increases, reaching the maximum value of 36%.

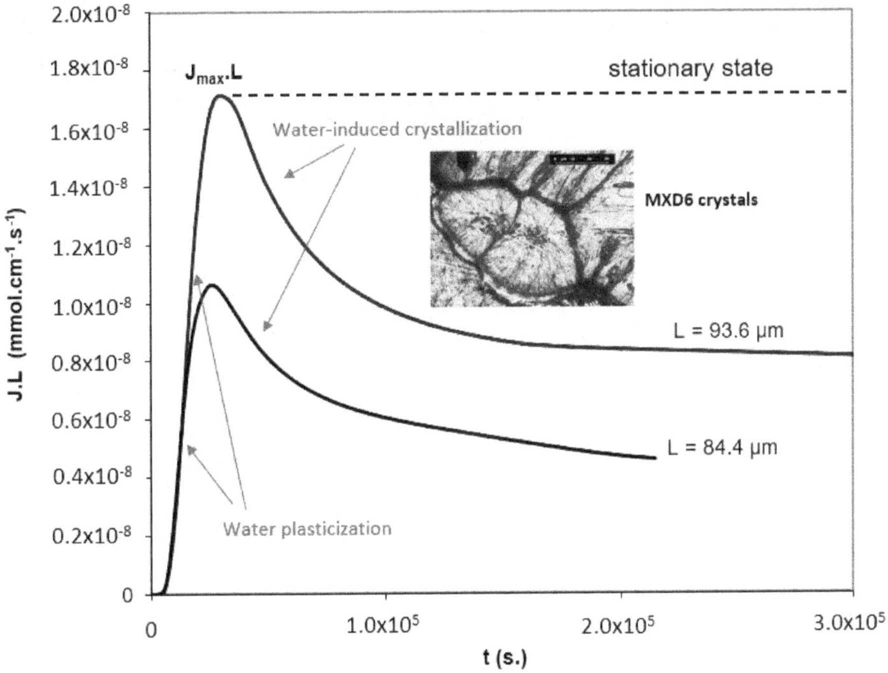

Figure 5.35 Water permeation flux through initially amorphous MXD6 films (of two different thickness L) at 25°C [97]

The water permeation fluxes through amorphous MXD6 films measured at 25 °C are unusual (Figure 5.35) as the crystals formed during the water permeation act as obstacle to the water migration, so that instead reaching a stationary state, the water flux is reduced progressively. Such phenomenon associated to the water plasticization effect led to complex diffusion mechanisms as a concentration and time dependent-diffusion coefficient was involved during the permeation process. Owing to this double dependence D = f(concentration, time), a deviation from the Fickian model should be considered and, so, the diffusion coefficients cannot be determined on the basis of simple Fick's laws.

It is interesting to note that the water concentration level in the polymer also influences the water permeation. Indeed, as the water-induced crystallization directly depends on the ability of the polymer chains to rearrange, so the swelling effect caused by water is linked to the polymer chains mobility. Besides, the swelling is as high as the absorbed water amount in the polymer is important. Therefore, the water sorption kinetics obtained in the case of MXD6 for various water vapor pressure (0–30 mbar) at 25°C highlight a decrease of the mass gain after reaching a maximum, but only at the highest water activity close to 1 [97, 98]. A BET II sigmoidal shape of the deduced water sorption isotherm is obtained, which is typical for the sorption behavior of glassy polymers and associated with the water aggregation (Figure 5.36). Thus, the decrease of the water uptake observed for the water permeation of MXD6 in contact with liquid water indicates the water-induced crystallization occurring when the water concentration is sufficiently high for strong swelling effect, i. e. when the MXD6 is in contact with saturated vapor pressure.

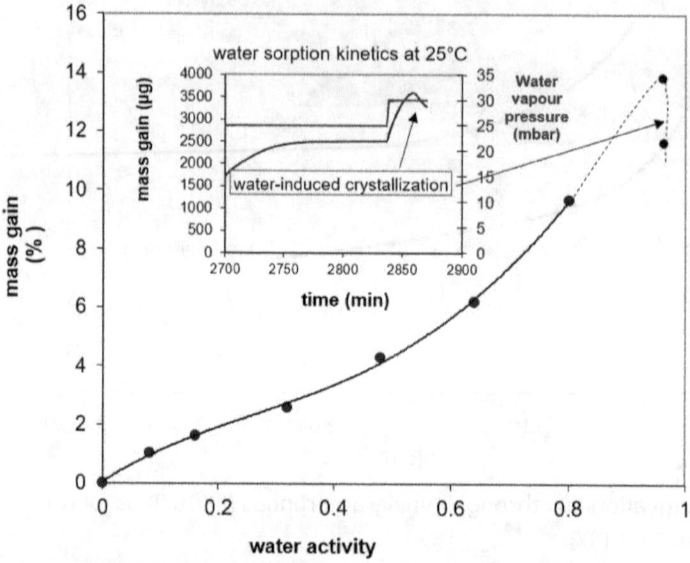

Figure 5.36 Water sorption isotherm of initially amorphous MXD6 at 25 °C; Insert: water sorption curve [98]

5.3.2.5 Aging – Degradation

The aging of a polymer material is considered to be a complex phenomenon corresponding to the slow and irreversible evolution of the material properties and resulting from the combined effects of interactions with the external environment, mechanical stress and the material instability [99]. In the case of thermoset resins, the instability can result from the not sufficiently crosslinked network that can release some residual compounds during the water absorption, as shown by Ellis and Found [100] by water sorption measurements performed on cured and postcured resins (Figure 5.37). It can be observed that the mass loss of a resin, which was postcured at 80 °C for 3 h (RT5), is much lower than for the same resin but cured at 22 °C for 18 h (RT4). The amount of extracted materials increases with increasing sorption temperature. Thus, the crosslinking degree of the three-dimensional network has a great influence on the sorption capacity of resins, and so on their stability in the long term.

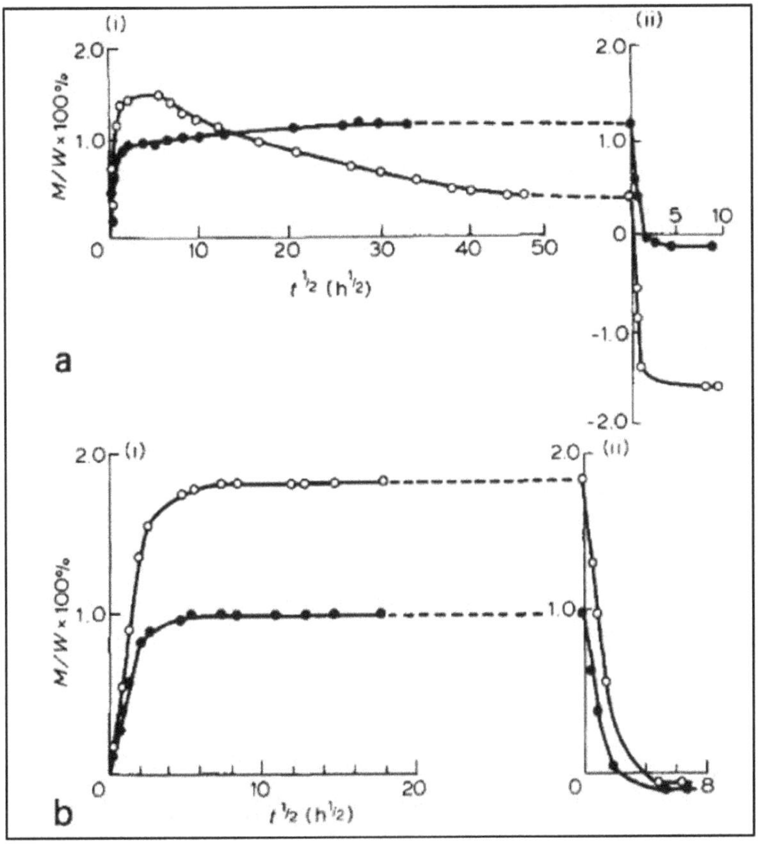

Figure 5.37 Liquid water sorption (i) and desorption (ii) kinetics of resins at 43.8 °C for: (a) room-temperature-cured resins, ● RT5 (film thickness of 386 µm), ○ RT4 (film thickness of 276 µm); (b) postcured resins, ● RT5 (film thickness of 532 µm), ○ RT4 (film thickness of 560 µm) [100]

Hydrolysis, which irreversibly affects the composition of the hydrophilic polymer, is one of the main chemical aging processes together with thermal aging or thermolysis (i. e. the decomposition of the polymer by heat) or thermo-oxidation (i. e. the polymer interaction with oxygen) or even radiolytic aging or radiolysis (i. e. the interaction of the material with the radiation). In the case of thermosetting network, the hydrolysis is caused by strong interactions of water molecules with the polar groups of the macromolecular network provoking the polymer chain scission, but also the crosslinking nodes breaking. Competition of the diffusion mechanism and reaction between the resin and the water occurs during hydrolysis, thus modifying the material behavior in water (namely, diffusivity).

Insufficient resin cross-linking during curing or incomplete post-curing will provide more hydrolyzable sites and, hence, more significant hydrolysis. In a polyester material, hydrolysis occurs according to the following reaction:

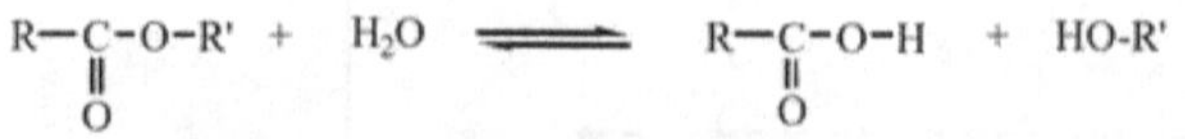

The breaking of macromolecular chains caused by the hydrolysis induces the decrease of the molecular mass and provokes the resin depolymerization [102]. The thermally activated hydrolysis can be catalyzed by various agents, such as OH^- or H^+. Indeed, the ester hydrolysis is faster in alkaline media due to the formation of the carboxylate anion, which is stabilized by the resonance structure and can react with alcohol groups. El Yagoubi et al. [103] studied the moisture sorption–desorption of an anhydride-cured epoxy resin. In this case the water sorption kinetics is described by a coupling "diffusion-reaction" mechanism, which suggests the hydrolysis reaction of the epoxy function into the diol function during the water diffusion process, thus inducing changes of the polymer structure.

Simar [49] also shows that the "diffusion-reaction" model is suitable to explain the water absorption in epoxy-based composites (Figure 5.38). He confirms that the insufficient curing treatment of the resin can modify its sorption capacity owing to the presence of the hydrolyzable groups.

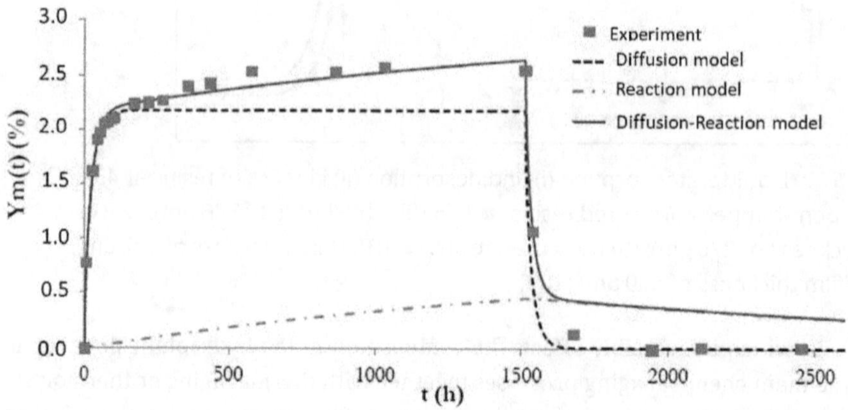

Figure 5.38 Experimental and calculated water sorption kinetics curves of epoxy resin at 70 °C and 85% RH [49]

The aging behavior of unsaturated polyester resin was extensively investigated by Richaud and Verdu [104]. In particular, the mechanisms of physical aging in the presence of solvent was studied. Kinetic curves with different shapes can be observed depending on the hydrolysis and/or leaching of unreacted chemicals during diffusion (Figure 5.18, curve 5, and Figure 5.39).

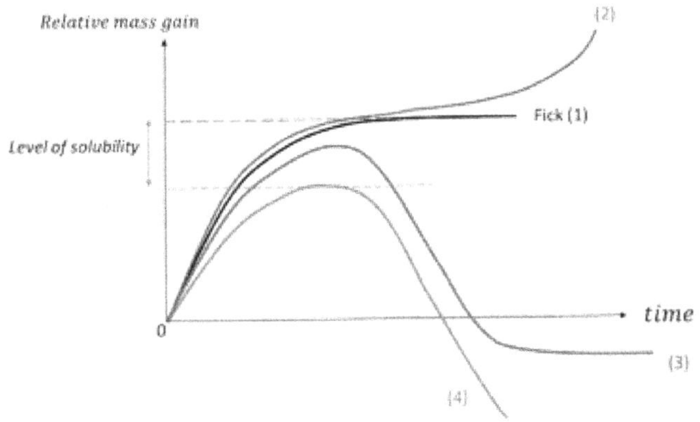

Figure 5.39 Theoretical kinetic curves: (1) Fickian diffusion; (2) water uptake followed by hydrolysis; (3) water uptake coupled with leaching of unreacted chemicals; and (4) water uptake followed by hydrolysis and extraction of hydrolysis by-products [104]

However, the difference between the curve shapes obtained during the water uptake coupled with leaching of unreacted chemicals (Figure 5.39, curve 3) and the water uptake followed by hydrolysis and extraction of hydrolysis by-products (Figure 5.39, curve 4) is not clear, as a relatively high content of initiator and catalyst is used, and can therefore be extracted by surrounding water molecules. Also, the determination of the real equilibrium mass gain is rather difficult, as other mechanisms can interfere, especially the hydrolysis (i. e. high mass loss after long immersion duration) at elevated temperatures.

The water cluster formation in polycarbonate was evidenced by Bair et al. [73] by sorption experiments performed at 97 °C. The water absorption in two stages was obtained: initially, all water molecules were found to be well-dispersed (i. e. non-aggregated) when cooled to room temperature, whereas the additional absorbed water was found during the second stage in a separate liquid phase corresponding to clustered water. The authors showed that below T_g the clusters are formed only after hydrolytic degradation of the polycarbonate (i. e. T_g decreases because of the decreased molecular mass caused by the hydrolysis), while above T_g clusters are formed without degradation. The concentration of clustered water was deduced by DSC analysis as the heat of water melting, and this value was 29% of the total water.

In the case of the polyester hydrolytic aging, Bellenger et al. [105] estimated the lifetime. Assuming a first-order hydrolytic process, the number of chain scissions n_t per mass unit at time t is given by:

$$\frac{dE}{dt} = \frac{dn_t}{dt} = k(E_0 - n_t)C_w \tag{5.46}$$

Where:

k = rate constant of hydrolysis

E_0 = initial concentration of ester groups (hydrolyzable)

C_w = water concentration

The increase of the material hydrophilic character (resulting from the ester hydrolysis into much more hydrophilic acid and alcohol groups) leads to the water uptake increasing. Thus, considering that the water equilibrium concentration is an additive function ($C_w = an_t + b$), Equation 5.46 can be expressed as follows:

$$\frac{dn_t}{dt} = k(E_0 - n_t)(b + an_t) \tag{5.47}$$

The solving of this equation leads to

$$n_t = E_0\left[1 - \frac{aE_0}{be^{[k(aE_0+b)t]} + aE_0}\right] \tag{5.48}$$

When hydrolysis kinetics are controlled by diffusion, Bellenguer et al. [105] show that for low conversion rate the "diffusion-reaction" coupling can be defined by the characteristic diffusion time τ_D and by the characteristic hydrolysis time τ_H as follows:

$$\tau_D = \frac{1}{k'} \text{ and } \tau_H = \frac{L^2}{D} \tag{5.49}$$

Where:

k' = rate constant

L = film thickness

D = water diffusion coefficient

When $\tau_H \ggg \tau_D$, water reaches the material bulk before it reacts and the degradation is homogenous, while when $\tau_H \lll \tau_D$, water reacts totally in the superficial layer and will never reach the bulk of the material. The degradation is heterogenous.

In contrast to hydrolytic aging, physical aging in some cases can reinforce the polymer moisture resistance owing to the material densification over time. For example, such an effect is observed for polycarbonate films after thermal treatment at 120 °C for different durations (0–2000 h) through the water permeation measurements [106]. A slight decrease of the water permeability was obtained after the aging, with a permeability of 1150 barrer[1)] obtained for the unaged film and 1040 barrer for the polycarbonate film aged at 120 °C for 2000 h. This result was explained by the polycarbonate densification as T_g increases and ΔC_p decreases, both being measured by DSC analysis.

The densification induced by the water penetration was highlighted by Garden and Pethrick [107] in the case of epoxy resins. The influence of the epoxy resin formulation on the moisture concentration was studied for samples cured at 40, 50, 60, and 70 °C and aged at 50 °C and 70 °C. It is shown that the water uptake depends on the difference between the glass transition and aging temperature. From Table 5.2, it can be seen that as the aging temperature increases, the water diffusivity for both systems increases, while an increase in both T_c and T_g causes a decrease in water uptake. The water uptake is lower for systems aged at the higher temperature (i. e. at 70 °C), whatever the system is. Thus, it is deduced that water initially sorbed in microvoids is re-dispersed with aging into the resin matrix allowing the plasticization and, subsequently, matrix densification. Moreover, the polymer chemical function groups influence strongly the polymer changes, and densification leads to a fall in the amount of absorbed water.

Table 5.2 Water Uptake and Diffusion Coefficients of Two Epoxy Systems (A and B) as a Function of the Curing Temperature T_c and Glass Transition Temperature T_g [107]

T_c (°C)	T_g (°C)	Aged at 50 °C		Aged at 70 °C	
		Water uptake (%)	D (10^9 cm^2·s^{-1})	Water uptake (%)	D (10^9 cm^2·s^{-1})
Epoxy A					
50	60	2.4	14.4 ± 0.8	2.7	63.5 ± 1.1
60	66	2.3	16.7 ± 1.0	2.3	63.0 ± 0.1
70	67	2.2	19.9 ± 1.2	2.2	63.6 ± 0.4
Epoxy B					
50	71	3.3	10.7 ± 0.4	2.8	45.0 ± 2.3
60	74	3.2	10.5 ± 0.6	2.7	44.3 ± 2.9
70	80	2.7	10.2 ± 0.8	2.6	38.9 ± 2.9

In ion exchange systems, such as Dowex resin and Nafion® 117, the aging effect on the water sorption has been studied by Ramkumar and Mulherjee [108]. The obtained results reveal the changes of the polymer physical structure due to long-storage or aging, thus decreasing the amount of sorbed water. This result is attributed to the fact that during long storage, the material loses its capacity to swell in water and the elas-

[1] 1 barrer = 10^{-10} cm$^3_{STP}$ cm cm^{-2} s^{-1} cm$_{Hg}^{-1}$

tic characteristics became rigid. So, the polymer matrix is hardened, and for Nafion® some kind of cross-linking can take place.

It is clear that the polymer moisture resistance can be modified by the aging; however, depending on the aging type, opposite effects can be sometimes observed. For example, moisture resistance *decreases* over time for hydrolyzable polymers, while polymers subject to age-induced densification by chain rearrangement show *increased* moisture resistance. In the worst case (i. e. when the degradation occurs after hydrolysis), water absorption promotes compound extraction, and so weakens the material's properties by creating microvoids and defects in the structure.

5.4 Conclusion

Due to their adaptative properties, processability, lightness, and relatively low cost, polymer materials are present in almost all industrial fields, such as automotive, aeronautics, energy, construction (buildings, roads, etc.), coatings, cosmetics, medical or even biomedical, but also packaging and in many products for domestic use. They are used on a large scale and are essential for everyday life. Over many years, the shortage of oil resources, the security of their supply, environmental issues linked to plastic waste, and finally industrial competition to lighten these materials while increasing their performance, have motivated those in research and development to find new strategies for choosing polymers, developing new materials (i. e. more complex structures combining several materials), and also for recycling and the handling a material's end of life.

The new technological challenges require polymer systems that will be more and more efficient in terms of thermal, mechanical, conductive, separative and barrier properties. Therefore, achieving gas and water resistance with the possibility of lightening the structure, are major challenges for manufacturers. The polymer "barrier" function is essential to:

- Preserve the quality of its contents (food, active ingredient, etc.)
- Guarantee the selectivity of membrane systems (for separation processes, filtration, energy production, etc.)
- Protect the external environment (for example, reservoirs, tanks for toxic species, conveyors, pipes, etc.)
- Ensure the operation of equipment sensitive to humidity and oxygen (i. e. biomedical devices, electronic, flexible screen (OLED), etc.).

The material stability depends on its barrier properties, owing to the aging and risk of migration that can provoke significant material and human damage during long-term applications. So, the challenge to obtain high barrier materials consists also in satisfying requirements for material flexibility, sizing, biocompatibility, and harmlessness (toxicological and microbiological).

During their application and use, polymers are often in contact with vapors of gases, organic compounds, and water. Thus, controlling their behavior towards these molecules and, in particular, towards water, is essential. Indeed, as is shown in this chapter, the water sorption in polymers initiates various phenomena that can affect the polymer transport properties, as well as its structure and morphology. Although moisture resistance is usually needed, some water sorption capacity can be required for some specific applications, such as fuel cells, water separation process, and absorbers.

The water sorption behavior of polymers has been well-studied in the literature, especially for dense polymers (namely thermoset resins), for which the moisture resistance is of prime importance and the mechanical performance is required over a long period.

Due to specific water–water and water–polymer interactions that appear during water penetration, the diffusive mechanism can be complex, owing to the concentration (plasticization), time (relaxation, aging), and concomitant phenomena induced by the swelling effect, which are all highly dependent on the external conditions of temperature and vapor pressure.

Because of the increasingly severe environmental constraints and competitive fields, new technologies should be developed in order to obtain high-performance materials with controlled transport properties. To face these challenges, a better understanding of the polymer properties in relation to their structure, morphology, and processing is essential for the prediction of the polymer behavior during long-term use. This is a key point for the design of more suitable materials with complex structures (i. e. composite, hybrid, or multinanolayered).

Deep analysis of water sorption in polymers, both in the kinetic and equilibrium (sorption isotherm) regimes, and over the whole range of the water activity (i. e. water in vapor and liquid states), is essential to better understand a polymer's behavior and its limitations towards the external environment. In addition to water, the transport of other diffusing species (i. e. gas and vapors) acting as probes at a molecular scale should be also studied for complete investigations of the material structural and transport properties.

To establish structure–property relationships, it is essential to combine permeation and sorption measurements with microstructural analysis, and also to study the aging and relaxation phenomena. Another key point is the material thickness, since the diffusion mechanism is usually studied on the lab-time scale (i. e. low thickness), while material thicknesses in practice are substantially higher. Therefore, the analysis of the structure-dependent thickness should be correlated, if possible, with the concentration profiles of the diffusing molecule. In general, structure–property relationships and the design of new materials with adaptive performances may be validated by comparing the experimental results with modelled data.

While real advances have been made in understanding solvent adsorption in polymers due to high-resolution measuring tools, there are still many questions about the sorption behavior of polymers under severe conditions of high temperature and high pressure, and for which the high performance is required for long-term applications, such as in energy (pipelines, tanks, etc.) and transport sectors. For these investigations, new and more suitable measuring devices will be necessary, which could bring new data for supporting predictive analysis for simulations with molecular dynamics and Monte Carlo methods.

References

[1] Modesti, M.; Dall'Acqua, C.; Lorenzetti, A.; Florian, E. Mathematical model and experimental validation of water cluster influence upon vapour permeation through hydrophilic dense membrane. *Journal of Membrane Science* 2004, 229, 211–223

[2] Crank, J.; Park G. S. *Diffusion in Polymers*, Academic Press, 1968

[3] Carslaw, H. S.; Jaeger J. C. *Conduction of Heat in Solids* (2nd edition), Oxford University Press, ch. III and XII, 1959

[4] Frisch, H. L. The time lag in diffusion. *Journal of Physical Chemistry* 1957, 61, 93–95

[5] Prager, S.; Long, F. A. Diffusion of hydrocarbons in polyisobutylene. *Journal of the American Chemical Society* 1951, 73, 4072–4075

[6] Marais, S.; Métayer, M.; Nguyen, Q. T.; Labbé, M.; Langevin, D. New methods for the determination of the parameters of a concentration-dependent diffusion law for molecular penetrants from transient permeation or sorption data. Macromolecular Theory and Simulations 2000, 9,. 207–214

[7] Follain, N.; Valleton J.-M., Lebrun, L.; Alexandre, B.; Schaetzel, P.; Métayer, M.; Marais, S. Simulation of kinetic curves in mass tranfer phenomena for a concentration-dependent diffusion coefficient in polymer membranes. Journal of Membrane Science 2010, 349, 195–207

[8] Kholer, R.; Dück, R.; Ausperger, B.; Alex, R. A numeric model for the kinetics of water vapor sorption on cellulosic reinforcement fibers. Composite Interfaces 2003, 10, 255–276

[9] Morton, W. E.; Hearle, J. W. S. *Physical Properties of Textile Fibers*, The Textile Institute, 1997

[10] Van Krevelen, D. W. Properties determining mass transfer in polymeric systems (ch. 18), in *Properties of Polymers* (3rd edition), Elsevier, 1990

[11] Stern, S. A.; Trohalaki, S. Fundamentals of gas diffusion in rubbery and glassy polymers, in *Barrier Polymers and Stuctures*, American Chemical Society, 1990

[12] Schaetzel, P.; Thomas, S.; Louahlia Gualous, H. Analogy between adsorption and sorption: an elementary mechanistic approach. I. Monolayer adsorption and sorption without solvent cluster formation. *Chinese Journal of Chemical Engineering* 2017, 25, 1740–1749

[13] Hirata, Y.; Marais, S.; Nguyen, Q. T.; Métayer, M.; Driouich, A.; Chevalier, L. Sorption, diffusion and permeation properties of oxygen and water in copolymer of ethylene and polar monomers. *Macromolecular Symposia* 2002, 188, 143–154

[14] Jovanović, D. S. Physical adsorption of gases. I: Isotherms for monolayer and multilayer adsorption, *Kolloid-Zeitschrift und Zeitschrift für Polymere* 1969, 235, 1203–1213

[15] Bungay, P. M.; Lonsdale, H. K.; De Pinho, M. N. *Synthetic Membranes: Science, Engineering and Applications*, Springer, 1986

[16] Favre, E.; Schaetzel, P.; Nguygen, Q. T.; Clément, R.; Néel, J. Sorption, diffusion and vapor permeation of various penetrants through dense poly(dimethylsiloxane) membranes: A transport analysis. *Journal of Membrane Science* 1994, 92, 169–184

[17] Berens, A. R. The solubility of vinyl chloride in poly(vinyl chloride). *Die Angewandte Makromolekulare Chemie* 1975, 47, 97–110

[18] Huggins, M. L. Some properties of long-chain compounds. *Journal of Physical Chemistry* 1942, 46, 151–158

[19] Jaroniec, M. Statistical interpretation of the Jovanović adsorption isotherms. *Colloid and Polymer Science* 1976, 254, 601–605

[20] Vopička, O.; Randová, A.; Friess, K. Sorption of vapours and liquids in PDMS: Novel data and analysis with the GAB model of multilayer adsorption. *European Polymer Journal* 2014, 60, 49–57

[21] Qian, C.; Mumby, S. J.; Eichinger, B. E. Phase diagrams of binary polymer solutions and blends. *Macromolecules* 1991, 24, 1655–1661

[22] Perrin, L.; Nguyen, Q. T.; Sacco, D.; Lochon, P. Experimental studies and modelling of sorption and diffusion of water and alcohols in cellulose acetate. *Polymer International* 1997, 42, 9–16

[23] Barrie, J.; Platt, B. The diffusion and clustering of water vapour in polymers. *Polymer* 1963, 4, 303–313

[24] Favre, E.; Nguyen, Q. T.; Clément, R.; Néel, J. The engaged species induced clustering (ENSIC) model: A unified mechanistic approach of sorption phenomena in polymers. *Journal of Membrane Science* 1996, 117, 227–236

[25] Singh, P. C.; Singh, R. K. Application of GAB model for water sorption isotherms of food products. *Journal of Food Processing and Preservation* 1996, 20, 203–220

[26] Timmermann, E. Q.; Chirife, J.; Iglesias, H. A. Water sorption isotherms of foods and foodstuffs: BET or GAB parameters? *Journal of Food Engineering* 2001, 48, 19–31

[27] Feng, H. Modeling of vapor sorption in glassy polymers using a new dual mode sorption model based on multilayer sorption theory. *Polymer* 2007, 48, 2988–3002

[28] Li, Y.; Nguyen, Q. T.; Fatyeyeva, K.; Marais, S. Water sorption behavior in different aromatic ionomer composites analyzed with a "new dual-mode sorption" model. *Macromolecules* 2014, 47, 6331–6342

[29] Watt, I. C. Adsorption–desorption hysteresis in polymers. *Journal of Macromolecular Science, Part A* 1980, 14, 245–255

[30] White, H. J.; Eyring, H. The adsorption of water by swelling high polymeric materials. *Textile Research Journal* 1947, 17, 523–553

[31] Giacobazzi, G.; Gioia, C.; Vannini, M.; Marchese, P.; Guillard, V.; Angellier-Coussy, H.; Celli, A. Water vapor sorption and diffusivity in bio-based poly(ethylene vanillate) – PEV. *Polymers* 2021, 13, 524

[32] Chen, M.; Coasne, B.; Guyer, R.; Derome, D.; Carmeliet, J. Role of hydrogen bonding in hysteresis observed in sorption-induced swelling of soft nanoporous polymers. *Nature Communications* 2018, 9, 3507

[33] Watt, I. C.; D'Arcy, R. L. Hydration of biopolymers. *Journal of Polymer Science: Polymer Symposia* 1976, 55, 155–166

[34] Salmen, L.; Larsson, P. A. On the origin of sorption hysteresis in cellulosic materials. *Carbohydrate Polymers* 2018, 182, 15–20

[35] Rivin, D.; Kendrick, C. E.; Gibson, P. W.; Schneider, N. S. Solubility and transport behaviour of water and alcohols in Nafion™. *Polymer* 2001, 42, 623–635

[36] Legras, M.; Hirata, Y.; Nguyen, Q. T.; Langevin, D.; MetayerM. Sorption and diffusion behaviors of water in Nafion 117 membranes with different counter ions. *Desalination* 2002, 147, 351–357

[37] Potreck, J.; Uyar, F.; Sijbesma, H.; Nijmeijer, K.; Stamatialis, D.; Wessling, M. Sorption induced reaxations during water diffusion in S-PEEK. *Physical Chemistry Chemical Physics* 2009, 11, 298–308

[38] Sgreccia, E.; Di Vona, M. L.; Licoccia, S.; Sganappa, M.; Casciola, M.; Chailan, J. F.; Auer, G.; Knauth, P. Self-assembled nanocomposite organic-inorganic proton conducting sulfonated poly-ether-ether-ketone (SPEEK)-based membranes: optimized mechanical, thermal and electrical properties. *Journal of Power Sources* 2009, 192, 353–359

[39] Hallinan, D. T.; Elabd, Y. A. Diffusion of water in nafion using time-resolved Fourier transform infrared-attenuated total reflectance spectroscopy. *Journal of Physical Chemistry B* 2009, 113, 4257–4266

[40] Berens, A. R. Effects of sample history, time, and temperature on the sorption of monomer vapor by PVC. *Journal of Macromolecular Science, Part B: Physics* 1977, 14, 483–498

[41] Wessling, M.; Lopez, M. L.; Strathmann, H. Accelerated plasticization of thin-film composite membranes used in gas separation. *Separation and Purification Technology* 2001, 24, 223–233

[42] Perrin, L. Etude de la diffusion et de l'équilibre de sorption de corps purs dans des polymères. Application à des systèmes utilisés en pervaporation (Ph.D. thesis), l'Ecole Nationale Supérieure des Industries Chimiques, 1992

[43] Moylan, C. R.; Best, M.; Ree, M. Solubility of water in polyimides: Quartz crystal microbalance measurements. *Journal of Polymer Science Part B: Polymer Physics* 1991, 29, 87–92

[44] Pranjoto, H.; Denton, D. D. Gravimetric measurement of steady-state moisture uptake in spin-coated polyimide film. *Journal of Applied Polymer Science* 1991, 42. 75–83

[45] Detallante, V.; Langevin, D.; Chappey, C.; Métayer, M.; Mercier, R.; Pinéri, M. Water vapor sorption in naphthalenic sulfonated polyimide membranes. *Journal of Membrane Science* 2001, 190, 227–241

[46] Fatyeyeva, K.; Rogalsky, S.; Makhno, S.; Tarasyuk, O.; Soto Puente, J. A.; Marais, S. Polyimide/ionic liquid composite membranes for middle and high temperature fuel cell application: Water sorption behavior and proton conductivity. *Membranes* 2020, 10, 82

[47] Follain, N.; Alexandre, B.; Chappey, C.; Colasse, L.; Médéric, P.; Marais, S. Barrier properties of polyamide 12/Montmorillonite nanocomposites: effect of clay structure and mixing conditions. *Composites Science and Technology* 2016, 136, 18–28

[48] Follain, N.; Crétois, R.; Lebrun, L.; Marais, S. Water sorption behaviour of two series of PHA/Montmorillonite films and determination of mean water cluster size. *Physical Chemistry Chemical Physics* 2016, 18, 20345–20356

[49] Simar, A. Impact du vieillissement humide sur le comportement d'un composite à matrice organique tissé fabriqué par injection RTM: Mise en évidence d'un couplage entre absorption d'eau et thermo-oxydation de la matrice (Ph.D. thesis), Ecole Supérieure de Mécanique d'Aérotechnique, Poitiers, 2014

[50] Ashbee, K.; Frank, F.; Wyatt, P. C. Water damage in polyester resins. *Proceedings of the Royal Society of London* 1967, 300, 415–419

[51] Langevin, D.; Grenet, J.; Saiter, J. M. Moisture sorption in PET. Influence on the thermokinetic parameters. *European Polymer Journal* 1994, 30, 339–345

[52] Domack, A.; Johannsmann, D. Plastification during sorption of polymeric thin films: A quartz resonator study. *Journal of Applied Physics* 1996, 80, 2599–2604

[53] Marais, S.; Métayer, M.; Labbé, M.; Legras, M. T.; Saiter J.-M. Water difusion in unsaturated polyester films. Effect of plasticization on the glass transition. *Polymer Engineering and Science* 1999, 39, 1508–1516

[54] Eisenberg, P.; Lucas, J. C.; Williams, R. J. J. Unsaturated polyesters: influence of the molar mass on the cure with styrene and the properties of the resulting network. *Journal of Applied Polymer Science* 1997, 65, 755–760

[55] Ríos De Anda, A. Influence of the solvent sorption, additivation, and chemical modification on the molecular mobility dynamics of polyamide 6,6 amorphous phase and its consequences on the tensile and impact strength properties of this polymer (Ph.D. thesis), University Claude Bernard Lyon I, 2012

[56] Cohen, M. H.; Turnbull, D. Molecular transport in liquids and glasses. *Journal of Chemical Physics* 1959, 31, 1164–1169

[57] Vrentas, J. S.; Duda, J. L.; Ling, H. C. Free-volume theories for self-diffusion in polymer-solvent systems. I. Conceptual differences in theories. *Journal of Polymer Science Part B: Polymer Physics* 1985, 23, 275–288

[58] Fujita, H. Diffusion in polymer-diluent systems. *Fortschritte der Hochpolymeren-Forschung* 1961, 3, 1–47

[59] Kokes, R. J.; Long, F. A. Diffusion of organic vapors into polyvinyl acetate. *Journal of the American Chemical Society* 1953, 75, 6142–6146

[60] Marais, S. Etude de la diffusivité de l'eau et de la solubilité de l'eau dans une résine polyester insaturé. Cinétiques de sorption et établissmeents des flux de perméation (Ph.D. thesis), Université de Rouen, 1998

[61] Marais, S.; Métayer, M.; Nguyen, Q. T.; Labbé, M.; Perrin, L.; Saiter J.-M. Permeametric and microgravimetric studies of sorption and diffusion of water in unsaturated polyesters. *Polymer* 2000, 41, 2667–2676

[62] Alix, S.; Philippe, E.; Bessadok, A.; Lebrun, L.; Morvan, C.; Marais, S. Effect of chemical treatments on water sorption and mechanical properties of flax fibres. *Bioresource Technology* 2009, 100, 4742–4749

[63] Chang, Y. P.; Karim, A. A.; Seow, C. C. Antiplasticization by water in reduced-moisture food systems. *Journal of Food Science* 1999, 64, 576–681

[64] Marzec, A.; Lewicki, P. P. Antiplasticization of cereal-based products by water. Part I. Extruded flat bread. *Journal of Food Engineering* 2006, 73, 1–8

[65] Marais, S.; Nguyen, Q. T.; Devallencourt, C.; Métayer, M.; Nguyen, T. U.; Schaetzel, P. Permeation of water through polar and nonpolar polymers and copolymers: determination of the concentration-dependent diffusion coefficient. *Journal of Polymer Science Part B: Polymer Physics* 2000, 38, 1998–2008

[66] Sabard, M. Etude de films PA6 et nanocomposites: Effet des conditions de procédé et des nanocharges sur la morphologie et les propriétés de sorption à l'eau et à l'éthanol (Ph.D. thesis), University of Lyon, 2011

[67] Picard, E.; Gérard J.-F., Espuche, E. Water transport prorertie of polyamide 6 based nanocomposites prepared by melt blending: on the importance of the clay dispersion state on the water transport properties at high water activity. *Journal of Membrane Science* 2008, 313, 284–295

[68] Starkweather, H. W. Some aspects of water clusters in polymers. *Macromolecules* 1975, 8, 476–479

[69] Lim, L. T.; Britt, I. J.; M. A. Tung, Sorption and transport of water vapor in Nylon 6,6 film. *Journal of Applied Polymer Science* 1999, 71, 197–206

[70] Sabard, M.; Gouanvé, F.; Espuche, E.; Fulchiron, R.; Seytre, G.; Fillot L.-A., Trouillet-Fonti, L. Influence of film processing conditions on the morphology of polyamide 6: Consequences on water and ethanol sorption properties. *Journal of Membrane Science* 2012, 415–416, 670–680

[71] Tenn, N.; Follain, N.; Soulestin, J.; Cretois, R.; Bourbigot, S.; Marais, S. Effect of nanoclay hydration on barrier properties of PLA/Monmorillonite based nanocomposites. *Journal of Physical Chemistry Part C* 2013, 117, 12117–12135

[72] Fatyeyeva, K.; Chappey, C.; Poncin-Epailard, F.; Langevin, D.; Valleton J.-M., Marais, S. Composite membranes based on Nafion® and plasma treated clay charges: Elaboration and water sorption investigations. *Journal of Membrane Science* 2011, 369, 155–166

[73] Bair, H. E.; Johnson, G. E.; Merriweather, R. Water sorption of polycarbonate and its effect on the polymer's dielectric behaviour. *Journal of Applied Physics* 1978, 49, 4976

[74] Zimm, B. H.; Lundberg, J. L. Sorption of vapors by high polymers. *Journal of Physical Chemistry* 1956, 60, 425–428

[75] Lundberg, J. L. Molecular clustering and segregation in sorption systems. *Pure and Applied Chemistry* 1972, 31, 261–281

[76] Strarkweather, H. W. Clustering of water in polymers. *Polymer Letters* 1963, 1, 133–138

[77] Aranda, P.; Chen, W. J.; Martin, C. R. Water transport across polystyrenesulfonate/alumina composite membranes. *Journal of Membrane Science* 1995, 99, 185–195

[78] Del Nobile, M. A.; Mensitiere, G.; Sommazzi, A. Gas and water vapor transport in a polyketone terpolymer. *Polymer* 1995, 36, 4943–4950

[79] Lozay, Q. Conception et caractérisation de matériaux composites nanostructurés à hautes propriétés barrières. Etude de films multinanocouches de PE et PA6 chargés de montmorillonite (Ph.D. thesis). University of Rouen, 2020

[80] Starkweather, H. W. Some aspects of water clusters in polymers. *Macromolecules* 1975, 8, 476–479

[81] Hunt, D. G. Prediction of sorption and diffusion of water vapour by Nylon-6,6. *Polymer* 1980, 21, 495–501

[82] Prykhodko, Y.; Fatyeyeva, K.; Hespel, L.; Marais, S. Progress in hybrid composite Nafion®-based membranes for proton exchange fuel cell application. *Chemical Engineering Journal* 2021, 409, 127329

[83] Mauritz, K. A.; Moore, R. B. State of understanding of Nafion. *Chemical Reviews* 2004, 104, 4535–4586

[84] Gierke, T. D.; Munn, G. E.; Wilson, F. C. The morphology in Nafion perfluorinated membrane products, as determined by wide- and small-angle X-ray studies. *Journal of Polymer Science: Polymer Physics Edition* 1981, 19, 1687–1704

[85] Hsu, W. Y.; Gierke, T. D. Ion transport and clustering in Nafion perfluorinated membranes. *Journal of Membrane Science* 1983, 13, 307–328

[86] Schmidt-Rohr, K.; Chen, Q. Parallel cylindrical water nanochannels in Nafion fuel-cell membranes. *Nature Materials* 2008, 7, 75–83

[87] Kreuer, K. D.; Paddison, S. J.; Spohr, E.; Schuster, M. Transport in proton conductors for fuel-cell applications: simulations, elementary reactions, and phenomenology. *Chemical Reviews* 2004, 104, 4637–4678

[88] Li, Y.; Nguyen, Q. T.; Lixon-Buquet, C.; Langevin, D.; Legras, M.; Marais, S. Water sorption in Nafion® membranes analyzed with an improved dual-mode sorption model – Structure/property relationships. *Journal of Membrane Science* 2013, 439, 1–11

[89] de Almeida, S. H.; Kawano, Y. J. Thermal behaviour of Nafion membranes. *Journal of Thermal Analysis and Calorimetry* 2004, 58, 569–577

[90] Chappey, C.; Fatyeyeva, K.; Rynkowska, E.; Kujawski, W.; Karpenko-Jereb, L.; Kelterer A.-M., Marais, S. Sulfonic membrane sorption and permeation properties: Complementary approaches to select a membrane for pervaporation. *Journal of Physical Chemistry Part B* 2017, 121, 8523–8538

[91] Berens, A. R.; Hopfenberg, H. B. Diffusion and relaxation in glassy polymer powders: 2. Separation of diffusion and relaxation parameters. *Polymer* 1978, 19, 489–496

[92] Chandra, P.; Koros, W. J. Sorption of lower alcohols in poly(ethylene terephthalate). *Polymer* 2009, 50, 4241–4249

[93] Ramani, R.; Ranganathaiah, C. Free-volume microprobe study of iodine diffusion in polymers. *Polymer International* 2001, 50, 237–248

[94] McDowell, C. C.; Freeman, B. D.; McNeely, G. W. Acetone sorption and uptake kinetic in poly(ethylene terephthalate). *Polymer* 1999, 40, 3487–3499

[95] Potreck, J.; Uyar, F.; Sijbesma, H.; Nijmeijer, K.; Stamatialis, D.; Wessling, M. Sorption induced relaxations during water diffusion in S-PEEK. *Physical Chemistry Chemical Physics* 2009, 11, 298–308

[96] Lixon, C.; Ben Doudou, B.; Chappey, C.; Dargent, E.; Marais, S. Permeation properties of poly (*m*-xylene adipamide) membranes. *Journal of Physical Chemistry Part B* 2009, 113, 3445–3452

[97] Ben Doudou, B.; Dargent, E.; Marais, S.; Grenet, J.; Hirata, Y. Barrier properties and microstructure modifications induced by liquid water for a semiaromatic polyamide. *Journal of Polymer Science Part B: Polymer Physics* 2005, 43, 2604–2616

[98] Ben Doudou, B. Etude du poly(*m*-xylène adipamide) et de ses dispersions dans une matrice poly(ethylène téréphtalate). Relations structures-propriétés (Ph.D. thesis). University of Rouen, 2005

[99] Verdu, J.; Fayolle, B. Vieillissement physique des matériaux polymères. *Techniques de l'Ingénieur* 2005, AM3150 01

[100] Ellis, B.; Found, M. S. The effects of water absorption on a polyester/chopped strand mat laminate. *Composites* 1983, 14, 237–243

[101] Apicella, A.; Migliaresi, C.; Nicolais l., Iaccarino l., Roccotelli, S. The water ageing of unsaturated polyester-based composites: influence of resin chemical structure. *Composites* 1983, 14, 387–392

[102] Jacquemet, R. Etude du comportement au vieillissement sous charges de stratifiés polyester/verre en milieu marin (Ph.D. thesis). University of Nancy I, 1989

[103] El Yagoubi, J.; Lubineau, G.; Roger, F.; Verdu, J. A fully coupled diffsuion-reaction scheme for moisture sorption–desorption in an anhydride-cured epoxy resin. *Polymer* 2012, 53, 5582–5595

[104] Richaud, E.; Verdu, J. in *Unsaturated Polyester Resins: Fundamentals, Design, Fabrication, and Applications*. Elsevier, 2019, ch. 9, pp. 99–231

[105] Bellenger, V.; Ganem, M.; Mortaigne, B.; Verdu, J. Lifetime prediction in the hydrolytic ageing of polyesters. *Polymer Degradation and Stability* 1995, 49, 91–97

[106] Diawara, B.; Fatyeyeva, K.; Ortiz, J.; Marais, S. Polycarbonate/mica extrusion using mixing elements: improvement of transparency and thermal, mechanical and water and gas barrier properties. *Polymer* 2021, 230, 124030

[107] Garden, L.; Pethrick, R. A. A dielectric study of water uptake in epoxy resin systems. *Journal of Applied Polymer Science* 2017, 134, 44717

[108] Ramkumar, J.; Mulherjee, T. Effect of aging on the water sorption and ion exchange studies on Nafion and Dowex resins: transition metal ion-proton exchange systems. *Separation and Purification Technology* 2007, 54, 61–70

Index

T

V

W

Y

www.ingramcontent.com/pod-product-compliance
Lightning Source LLC
LaVergne TN
LVHW081928200926
848484LV00009B/2392
* 9 7 8 1 5 6 9 9 0 7 9 3 1 *